Blood Gases and Electrolytes

Blood Gases and Electrolytes

Second Edition

John Toffaletti, PhD

1850 K Street, N.W., Suite 625
Washington, DC 20006-2213

© 2009 American Association for Clinical Chemistry, Inc. All rights reserved. No part of this publication may be reproduced, stored in a retrieval system, or transmitted in any form by electronic, mechanical, photocopying, or any other means without written permission of the publisher.

1 2 3 4 5 6 7 8 9 0   WPC   11 10 09

Printed in the United States of America

Library of Congress Cataloging-in-Publication Data

Toffaletti, John G.
  Blood gases and electrolytes / John G. Toffaletti. – 2nd ed.
    p. ; cm.
  Rev. ed. of: Special topics in diagnostic testing : blood gases and electrolytes. c2001.
  Includes bibliographical references and index.
  ISBN-13: 978-1-59425-097-2
  ISBN-10: 1-59425-097-9
  1. Blood gases–Analysis. 2. Water-electrolyte imbalances–Diagnosis. 3. Electrolytes. I. Toffaletti, John G. Special topics in diagnostic testing. II. American Association for Clinical Chemistry. III. Title.
  [DNLM: 1. Blood Gas Analysis. 2. Electrolytes. QY 450 T644b 2009]

  RB45.2.T64 2009
  616.07'561–dc22

2009012651

Contents

About the Author   viii

Chapter 1	Foundations of Blood Gases: Acid-Base Balance, Oxygenation Status, and Interpretation	1
	History of Blood Gases	1
	Explanation of Blood Gas, Acid-Base, and Cooximetry Terms	2
	Physiology of Acids and Bases: How are Acids and Bases Produced?	8
	Buffer Systems	9
	Acid-Base Regulation	11
	Hemoglobin Binding	13
	Oxygen Delivery to Tissues	14
	Clinical Abnormalities of Acid-Base Balance	15
	Detecting Mixed Acid-Base Disorders	23
	Disorders of Oxygenation: Hypoxemia and Tissue Hypoxia	26
	Evaluation of Blood Gas Results	26
	Evaluation of Oxygenation Status	31
	Collection and Handling of Samples for Blood Gas Analysis	35
	Cord Blood Gases	36
	References	37
Chapter 2	Calcium, Magnesium, and Phosphate	41
	History and Significance	41
	Calcium	42
	Physiology	42
	Regulation in the Blood	43
	Distribution in Cells and Blood	45
	Evaluation of Calcium Measurements	46
	Causes of Hypocalcemia	46
	Causes of Hypercalcemia	50
	Interpretation of Calcium and PTH Measurements	52
	Proper Collection and Handling of Samples	53
	Reference Ranges for Calcium	54

	Magnesium	54
	Physiology	54
	Regulation in the Blood	55
	Distribution in Cells and Blood	56
	Evaluation of Magnesium Status	57
	Causes of Hypomagnesemia	57
	Treatment of Hypomagnesemia	63
	Causes of Hypermagnesemia	63
	Proper Collection and Handling of Samples	64
	Diagnosis of Magnesium Deficiency	64
	Reference Ranges for Magnesium	65
	Phosphate	66
	Physiology	66
	Regulation in Blood	66
	Distribution in Cells and Blood	66
	Evaluation of Hypophosphatemia	67
	Causes of Hypophosphatemia	68
	Symptoms of Hypophosphatemia	69
	Treatment of Hypophosphatemia	69
	Evaluation of Hyperphosphatemia	69
	Causes of Hyperphosphatemia	70
	Clinical Consequences of Hyperphosphatemia	71
	Treatment of Hyperphosphatemia	71
	Reference Ranges for Phosphate	71
	References	72
Chapter 3	Electrolytes: Sodium, Potassium, Chloride, and Bicarbonate	77
	Osmolality and Volume Regulation	77
	Physiology	77
	Regulation of Osmolality	78
	Regulation of Blood Volume	79
	Reference Ranges for Osmolality	81
	Sodium	81
	Physiology and Regulation	81
	Evaluation of Hyponatremia	82
	Treatment of Hyponatremia	86
	Evaluation of Hypernatremia	86
	Treatment of Hypernatremia	88
	Reference Ranges for Sodium	88
	Potassium	88
	Physiology	88
	Regulation	89
	Causes of Hypokalemia	90

	Clinical Effects of Hypokalemia	91
	Treatment of Hypokalemia	91
	Causes of Hyperkalemia	92
	Clinical Effects of Hyperkalemia	93
	Treatment of Hyperkalemia	94
	Proper Collection and Handling of Samples	94
	Reference Ranges for Potassium	94
Chloride		95
	Physiology and Regulation	95
	Causes of Hypochloremia and Hyperchloremia	95
	Reference Ranges for Chloride	96
Bicarbonate		96
	Physiology and Regulation	96
	Causes of Decreased and Increased Bicarbonate	97
	Reference Ranges for Bicarbonate	97
References		97
Appendix	Self-Assessment and Mastery	99
Index		115

About the Author

John Toffaletti, PhD, graduated with honors from the University of Florida in Gainesville with a bachelor of science degree in chemistry. He then trained in clinical chemistry at the University of North Carolina at Chapel Hill, where he earned a doctoral degree in biochemistry with Drs. John Savory and Hill Gitelman. His postdoctoral fellowship in clinical chemistry followed, at Hartford Hospital with Dr. George Bowers.

After completing his training, Dr. Toffaletti began working in the clinical laboratories at Duke University Medical Center, where he has been since 1979 and where he currently serves as Professor of Pathology, Director of the Blood Gas Laboratory, the Clinical Pediatric Laboratory, several Outpatient Laboratories, and Associate Director of Clinical Chemistry. He is also the Director of Clinical Chemistry at the Durham VA Medical Center.

As a member of AACC, Dr. Toffaletti served as Chairman of the Contributed Papers Committee for the 1984 and 1997 Annual Meetings, Chairman of the North Carolina Section in 1983–4, Chairman of the Clinical Chemistry News Board of Editors in 1990, Chairman of the Electrolyte/Blood Gas Division in 1991 and 1992, and Chairman of the Commission on Publications in 1993. From 1999 to the present, he has served on the Board of Editors of *Clinica Chimica Acta*. Most recently, he chaired the Scientific Program Committee for the 2006 Critical Care and Point of Care Symposium in Quebec.

The material in this book is derived from the numerous workshops, seminars, study guides, and book chapters that Dr. Toffaletti has created about the interpretation of blood gas, co-oximetry, ionized calcium, magnesium, lactate, and renal function tests. His research interests include sample collection, analysis, and clinical use of these tests.

Chapter 1

Foundations of Blood Gases: Acid-Base Balance, Oxygenation Status, and Interpretation

"Blood gases" refers to the parameters pH, pCO_2, and pO_2, which are commonly measured in blood. Note that the lower-case "p" in pH stands for negative log, while the italicized p in pCO_2 and pO_2 stands for the partial pressure of each of these gases. Blood gases are commonly measured by electrochemical (potentiometric or amperometric) methods that use ion-selective or gas-selective electrodes. Here, we use the unit mmHg for blood gases because the unit is readily understood and because the many journals that follow the American Medical Association's style guidelines use mmHg as the conventional and SI unit.

"Oximetry" refers to the measurement of various forms of hemoglobin (Hb), including oxyhemoglobin (O_2Hb), from which the term is derived. Oximeters are specialized spectrophotometers that measure the absorbance of multiple wavelengths to calculate the various Hbs present in blood.

This chapter presents basic foundational knowledge of blood gases with respect to acid-base balance, oxygenation status, and interpretation. Included are a history of the study of blood gases; an explanation of terms; a description of blood gas physiology, as well as related disorders and their evaluation; and recommended techniques for collection and handling of samples for blood gas analysis.

HISTORY OF BLOOD GASES

The history of blood gases and oximetry has perhaps the oldest, best documented, and, to some of us, the most interesting history of developments in laboratory tests. The history includes Joseph Priestley, a clergyman and natural philosopher, who became fascinated with gases by observing the large volume of gases produced in making beer, then went on to isolate 10 gases, including oxygen. There is Henry "Hank" Cavendish, a nerd even by the standards of 200 years ago, who, in addition to discovering hydrogen, unexpectedly inherited the modern equivalent of about a billion dollars. The early history of blood gases even includes Benjamin Franklin, a colleague of many scientists, including Priestley. To paraphrase Alan Grogono,

In addition to publishing newspapers, drafting constitutions, serving as postmaster general, flying kites in thunderstorms, discovering the Gulf Stream, and maintaining friendships with French ladies, Benjamin Franklin found time to make an unfortunate guess about calling "vitreous" charges "positive."

This decision led to assigning a "negative" charge to electrons and a "positive" charge to hydrogen ions *(1)*.

More recently, we have the acid-base work of notables such as Lawrence Henderson and Karl Hasselbalch, who developed their famous equation, but who never knew each other. Several of their predecessors and contemporaries had tried to define an acid. Arrhenius, for example, defined acids as hydrogen salts. Brønsted and Lowry simultaneously, but separately, defined acids as substances that could donate a hydrogen ion, and Lowry later added that an acid could accept a pair of electrons to form a covalent bond. Donald Van Slyke embraced the idea that acid-base status was partly determined by electrolytes, an idea that was expanded by Peter Stewart into the very complex "strong ion difference" explanation of acid-base balance *(2)*. Van Slyke is also credited with expanding chemical analyses into the hospital and is considered a founder of clinical chemistry. Most notably, around 1920, he developed a "gasometric" method (measured released O_2 gas) for measuring oxygen saturation in blood.

Before and during World War II there was the work by Kurt Kramer, J.R. Squires, and Glen Millikan on oximeters, especially those for use in oxygen delivery systems for high-altitude military flights. (These developments eventually led to the discovery of pulse oximetry by Takuo Aoyagi in 1975.) Later, in 1954, Leland Clark, using polyethylene film and other materials that cost less than a dollar, developed the prototype electrode for partial pressure of oxygen (pO_2). Also in 1954, Richard Stowe covered a pH electrode with a rubber finger cot to develop a prototype of today's partial pressure of carbon dioxide (pCO_2) electrodes. These stories and many others were documented in a book by Astrup and Severinghaus *(3)*.

EXPLANATION OF BLOOD GAS, ACID-BASE, AND COOXIMETRY TERMS

pH

pH is an index of blood acidity or alkalinity. If normal arterial pH is 7.35–7.45, then a pH <7.35 indicates an acid state, and a pH >7.45 indicates an alkaline state. In critical care, a clinically acceptable range of 7.30–7.50 is sometimes used as a guideline (see Table 1-1). Acidemia means the blood is too acidic, and acidosis refers to the metabolic process within the patient that causes the acidemia. (The adjective for the process is acidotic.) Similar terms are used for the alkaline state: alkalemia, alkalosis, and alkalotic. Because all enzymes and physiological processes may be affected by pH, pH is normally regulated within close tolerances.

Foundations of Blood Gases

TABLE 1-1. Reference Ranges for Venous and Arterial Blood

Measurement	Reference range (mixed venous)	Reference range (arterial)	Clinically acceptable range (arterial)
pH	7.33–7.43	7.35–7.45	7.30–7.50
pCO_2	41–51 mmHg (5.5–6.8 kPa)	35–45 mmHg (4.7–6.0 kPa)	30–50 mmHg (4.0–6.7 kPa)
HCO_3^-	21–30 mmol/L (21–30 mEq/L)	21–28 mmol/L (21–28 mEq/L)	Variable
pO_2	35–40 mmHg (4.7–5.3 kPa)	83–108 mmHg (11.1–14.4 kPa^a)	>80 mmHg (>10.7 kPa^a)
sO_2 (%)	70–75	96–100	>90
$\%O_2Hb$	—	94–99	≤90
Anion gap	8–16 mmol/L	Same	—
Base excess	−3 to +3	Same	—

$^a pO_2$ of arterial blood varies with age.

pCO_2

pCO_2 is a measure of the tension or pressure of carbon dioxide dissolved in the blood. The pCO_2 of blood represents the balance between cellular production of CO_2 and ventilatory removal of CO_2. A normal, steady pCO_2 indicates that the lungs are removing CO_2 at about the same rate as tissues are producing CO_2. A change in pCO_2 indicates an alteration in this balance, usually due to a change in ventilatory status. CO_2 is an acidic gas that is largely controlled by our rate and depth of breathing or ventilation, provided that the rate of metabolic production of CO_2 is constant. pCO_2 is the respiratory or ventilatory component of acid-base balance.

pO_2

pO_2 is a measure of the tension or pressure of oxygen dissolved in the blood. The pO_2 of arterial blood is primarily related to the ability of the lungs to oxygenate blood from alveolar air. A decreased arterial pO_2 indicates one or more of the following circumstances:

- Decreased pulmonary ventilation, as caused, for example, by airway obstruction or trauma to the brain
- Impaired gas exchange between alveolar air and pulmonary capillary blood, as caused, for example, by bronchitis, emphysema, pulmonary edema, or asthma
- Altered blood flow within the heart or lungs, as caused by congenital defects in the heart or shunting of venous blood into the arterial system without oxygenation in the lungs

Bicarbonate

Although bicarbonate ion (HCO_3^-) can now be measured directly, most blood gas analyzers calculate HCO_3^- using the Henderson-Hasselbalch equation from measurements of the pH and pCO_2. Bicarbonate is an indicator of the buffering capacity of blood; a low bicarbonate indicates that a larger pH change will occur for a given amount of acid or base produced. Bicarbonate is classified as the metabolic component of acid-base balance.

Base Excess

Base excess (BE) is a calculated term that describes the amount of bicarbonate relative to pCO_2. Some believe it helps to quickly determine the amount of bicarbonate that a patient may need, and provides more useful information about acid-base status than the bicarbonate and pCO_2 (4).

The BE concept was introduced by Astrup and Siggaard-Andersen in 1958. The equation for calculating base excess (BE) from the HCO_3^- and pH is not at all intuitive, but it is based on the relationship between pH, pCO_2, and HCO_3^-. It also somehow includes the contribution of hemoglobin as a buffer (4).

$$BE = 0.929 [HCO_3^- - 24.4 + 14.8 (pH - 7.4)] \qquad \text{Eqn (1)}$$

Standard Base Excess (SBE) is the Base Excess (BE) value calculated for hypothetical anemic blood (Hb = 5 g/dL) on the principle that blood hemoglobin effectively buffers both the plasma and the much larger amount of extracellular fluid. It is as if the blood hemoglobin were dispersed in the larger pool of extracellular fluid. The SBE predicts the quantity of acid or alkali required to return the plasma in-vivo to a normal pH (7.35–7.45) with the amount of carbon dioxide held at a standard value. The normal reference range is −3 to +3 mmol/L. Comparison of the calculated BE to the reference range for BE may help determine whether an acid/base disturbance is a respiratory, metabolic, or mixed metabolic/respiratory problem.

A base excess value exceeding +3 indicates alkalosis, such that the patient requires increased amounts of acid to return the blood pH to neutral. A base excess below −3 indicates that the patient is acidotic, and excess acid must be removed from the blood to return the pH to normal.

Another definition for base excess is the amount of acid or base that must be added to a liter of blood (ECF) to return the pH to 7.4 at a pCO_2 of 40 mmHg.

Common clinical situations with a base excess less than −3 include production of lactate and acid from hypoxia and diabetic ketoacidosis. A common situation observed with a base excess exceeding +3 is persistent vomiting, causing loss of acidic gastric fluids.

Table 1-2 shows the relationship between BE and the simple difference between 24 mmol/L (a very normal bicarbonate) and the measured bicarbonate. I leave it to each reader to determine the clinical importance of calculating BE vs. use of the bicarbonate concentration.

TABLE 1-2. Comparisons of Base Excess to Bicarbonate Difference

pH	HCO_3^-	pCO_2	BE	$HCO_3^- - 24$
7.2	11	30	−14	−13
7.2	15	40	−11	−9
7.2	19	50	−8	−5
7.2	23	60	−4	−1
7.3	14	30	−11	−10
7.3	19	40	−6	−5
7.3	24	50	−2	0
7.4	18	30	−6	−6
7.4	24	40	−0.4	0
7.4	30	50	5	6
7.5	23	30	−0.3	−1
7.5	30	40	7	6
7.5	38	50	14	14
7.6	19	20	−2	−5
7.6	29	30	7	5
7.6	38	40	15	14

Anion gap (AG) is a calculated term that represents the difference between the commonly measured cations (sodium [Na] and sometimes potassium [K]) and the commonly measured anions (chloride [Cl] and bicarbonate [HCO_3^-]). Therefore, it represents the unmeasured anions such as lactate, acetoacetate, and albumin. The calculation is this:

$$AG = [Na^+] - [Cl^-] - [HCO_3^-] \text{ (reference range: 8–16 mmol/L)} \quad \text{Eqn (2)}$$

If K is included in the calculation, the formula changes to this:

$$AG = [Na^+] + [K^+] - [Cl^-] - [HCO_3^-]$$
$$\text{(reference range: 12–20 mmol/L)} \quad \text{Eqn (3)}$$

The AG is useful in diagnosing a metabolic acidosis and differentiating among the causes, as shown in Table 1-3. For example, in metabolic acidosis due to hypoxia, the blood lactate increases, and this increases the AG. While blood lactate has become a common measurement, it is still considered an "unmeasured" anion.

While often very useful, AG is calculated from three or four measurements and can vary up to ±4 mmol/L. In patients with an AG of 20–29 mmol/L, about two-thirds will have a metabolic acidosis, while all patients with an AG higher than this will have a metabolic acidosis *(5,6)*.

Strong ion difference (SID) is a concept developed by Peter A. Stewart in 1981, aimed at explaining pH changes and at assessing clinical acid-base disturbances *(2,7,8)*. He used electroneutrality, conservation of mass, and the degree of dissociation (strong or weak) of electrolytes to explain acid-base physiology. It is complex mathematically and not easily grasped chemically, but I will try

TABLE 1-3. Changes of Anion Gap in Various Acid-Base Disorders

Disorder	Decreased	Gained	Effect on AG
Diarrhea	HCO_3^-	Cl	Little
Renal tubular acidosis	HCO_3^-	Cl	Little
Lactate acidosis	HCO_3^-	Lactate	Increased
Ketoacidosis	HCO_3^-	Ketoacids	Increased
Mixed disorder: ketoacidosis with metabolic alkalosis	HCO_3^-	Ketoacids and HCO_3^-	Increased (with little change in total CO_2)

to clarify the concept. In the simplest terms that I can understand, the SID concept says that the concentrations of H^+, OH^-, HCO_3^-, and a variety of other weak acids and bases (and therefore the pH) depend on the difference between strongly dissociated cations (like Na and K) and anions (such as Cl and lactate), with a higher SID favoring an alkaline environment and a lower SID favoring an acidic environment. Here is an easily grasped equation for SID:

$$SID = [Na^+] + [K^+] - [Cl^-] - [lactate] \qquad \text{Eqn (4)}$$

All these ions are strongly dissociated. Note that neither H^+ nor HCO_3^- are strongly dissociated ions and that Stewart considered albumin the most important weak acid *(2)*. This approach helps explain how a decreased albumin is sometimes associated with a metabolic alkalosis. Story et al. *(7)* proposed some relatively simple (but still hard to grasp) equations to predict the Unmeasured Ion Effect (UIE) from the Standard Base Excess (SBE), the Na-Cl effect, and the albumin effect:

$$UIE \text{ (mmol/L)} = SBE - [Na - Cl - 38] - 0.25 \times [42 - \text{Albumin (g/L)}] \qquad \text{Eqn (5)}$$

One example of how the SID works is explaining how changes in Cl ion concentration cause significant pH changes in the fluids that Cl ions are either moving into or out of. If a Cl ion moves across a membrane, it must either take a cation with it or exchange itself for another anion. If the Cl ion takes another strongly dissociated ion with it, such as Na, there will be no net change in the SID and no change in pH. However, if Cl ion exchanges for HCO_3^- ion, the SID of the solution receiving the Cl ion will decrease and become more acidic. Obviously, the HCO_3^- ion will make that solution more alkaline. The SID theory adds nothing to explain these mechanisms.

However, suppose that *only* a Cl ion is transported. A possible example of this is the production of gastric acid secreted by the gastric mucosa, which is ~0.1 mol/L HCl (pH~1). Oxyntic cells from the gastric tissue produce this fluid by transporting Cl ions from the cells into the gastric fluid. According to the SID theory, in the solution receiving the Cl ion, H_2O would dissociate in the gastric fluid (to maintain electroneutrality), with H^+ remaining with the Cl^- and OH^- moving into the oxyntic cell. The gain of Cl^- and H^+ would decrease the SID and produce a very acidic gastric fluid *(7,8)*. The cell and then the blood become more alkaline, but because blood continually circulates and buffers pH changes, it becomes only mildly alkaline. This has been known for many years as the "alkaline tide" of blood following meals. Table 1-4 illustrates these changes in SID and pH for a hypothetical oxyntic cell and gastric fluid.

While the SID theory may correctly explain some actual ion movements and pH changes, I have yet to be convinced that the SID explains all transcellular ion movements and pH changes better than conventional mechanisms for ion movements.

Hemoglobin (Hb) and its derivatives are measured by cooximetry. These derivatives include oxyhemoglobin (O_2Hb), deoxyhemoglobin (HHb), carboxyhemoglobin (COHb), and methemoglobin (metHb). O_2Hb has four oxygen molecules bound to each of the heme groups in the hemoglobin molecule. COHb has carbon monoxide (CO) bound very tightly to an O_2-binding site on a heme group, which also causes the remaining O_2 molecules to bind very tightly to the other three heme groups. Hb cannot release O_2 until the CO is released. MetHb is inactive Hb (unable to bind O_2) because its ferrous ion (Fe^{2+}) has been oxidized to a ferric ion (Fe^{3+}). Hemoglobin function is discussed further in the section "Hemoglobin Binding."

sO_2 and %O_2Hb

sO_2 is the percent oxygen saturation, which is the percentage of functional Hb that is saturated with oxygen. It is calculated (see Eqn 6, below) as the concentration of O_2Hb divided by the functional Hb. Although sO_2 is clearly related to oxygenation in the lungs, pO_2 is somewhat better for assessing that function. sO_2 may vary from 0% to 100% *(2)*.

$$sO_2 = O_2Hb/(O_2Hb + HHb) \qquad \text{Eqn (6)}$$

The %O_2Hb (formerly called the fractional Hb saturation) is the percentage of total Hb that is saturated with oxygen. The %O_2Hb may be used for determining oxygen content and, therefore, oxygen delivery (DO_2) and oxygen consumption (VO_2) *(9)*. It is calculated as follows:

$$\%O_2Hb = O_2Hb/(O_2Hb + HHb + COHb + MetHb) \qquad \text{Eqn (7)}$$

TABLE 1-4. Examples of Strong Ion Differences

	Hypothetical oxyntic cell composition	Hypothetical gastric fluid content
Na (mmol/L)	100	100
K (mmol/L)	0	0
H^+ (mmol/L)	~0	10
Cl (mmol/L)	100	110
Lactate (mmol/L)	0	0
SID (mmol/L)	0	−10
pH	Neutral	Very acidic

The clinical usefulness and confusion between these two terms for oxygen bound to hemoglobin (sO_2 and $\%O_2Hb$) are discussed in a recent report *(10)*.

CO-Hb and Met-Hb

Carboxyhemoglobin (CO-Hb) and methemoglobin (met-Hb) together normally make up only 1–2% of the total Hb in blood. Neither is able to perform O_2 carrying and releasing functions. CO-Hb is Hb with carbon monoxide (CO) bound very tightly to the site which normally binds O_2. (O_2, CO, and NO share a similar molecular structure.) Exposure to carbon monoxide will increase the %CO-Hb from about 1% to 5–10% in smokers and to 50% or more with exposure to toxic or lethal levels. Methemoglobin is Hb that has its Fe^{2+} ion oxidized to an Fe^{3+}, which also inactivates the ability of met-Hb to bind O_2.

DO_2 and VO_2

Both DO_2 (oxygen delivery) and VO_2 (oxygen consumption) are parameters for assessing critically ill patients. There are now several ways of determining these parameters, with laboratory measurements sometimes used. DO_2 in mL/min requires measurement of the following values in arterial blood: Hb in g/L, $\%O_2Hb$ from 0 to 1.0, pO_2 in mmHg (of minor importance in this calculation), and cardiac output (CO) in L/min. The calculation is below:

$$DO_2 = CO(1.34 \times Hb \times \%O_2Hb + 0.03\, pO_2) = CO(O_2a) \qquad \text{Eqn (8)}$$

where O_2a = arterial oxygen content.

VO_2 is oxygen consumption by the organs and tissues of the body. Put simply, it is the difference in oxygen content between arterial (O_2a) and venous (O_2v) blood multiplied by CO:

$$VO_2 = CO(O_2a - O_2v) \qquad \text{Eqn (9)}$$

PHYSIOLOGY OF ACIDS AND BASES: HOW ARE ACIDS AND BASES PRODUCED?

"Metabolic acid"

The vast majority of our metabolic acid is actually CO_2 produced in the mitochondria as one of the ultimate byproducts of glucose oxidation. However, CO_2 is considered the "respiratory" component of our acid-base balance because lung ventilation directly affects the blood pCO_2.

"Lactate" acidosis

Contrary to common belief, lactic acid is virtually never produced. In fact, the only reaction in the human that produces lactate is when pyruvate is converted

Foundations of Blood Gases

to lactate. Far from producing acid, this reaction actually *consumes* acid by the participation of NADH being converted to NAD+ *(11)*.

So where does the acidosis come from during an oxygen deficit? It happens many biochemical steps later, when ATP is converted to ADP, phosphate, and hydrogen ions. Normally, oxidative phosphorylation reconverts these products back to ATP with little net change of H ions. However, without oxygen, the ADP and acid accumulate to cause acidosis.

Ketoacidosis

Diabetic ketoacidosis develops when a person has too little insulin, such that glucose can't enter cells for energy. The liver begins to break down fat for energy, which produces toxic ketoacids formed by the deamination of amino acids and the breakdown of fatty acids. The two common ketones produced in humans are acetoacetate and β-hydroxybutyrate.

Ketoacidosis can also occur in people undergoing starvation, as the body is forced to break down fat for sustenance due to their lack of outside nutrition. In ketoacidosis, unregulated ketone production causes a severe accumulation of keto acids in the blood.

Production of base

Generally, there is little metabolic production of bicarbonate or other alkaline substances. Bicarbonate is generated from CO_2 produced by metabolism after combination with H_2O and dissociation of H^+. The kidney can retain or excrete HCO_3^- depending on the physiologic need. Bicarbonate can increase if Cl is lost, either by vomiting or by renal loss from diuretics. Generally, any loss of acid will increase the proportion of HCO_3^- from CO_2. Bicarbonate given in excess to treat an acidosis is an iatrogenic source of base.

BUFFER SYSTEMS

Bicarbonate–carbon dioxide (carbonic acid)

This is the buffer in highest concentration (~24 mmol/L) in the blood plasma, which is also of central importance in acid buffering in the blood. CO_2 is a volatile acidic gas, is soluble in water, and is the major acid produced from energy metabolism. CO_2 produced by metabolism readily diffuses from cells into blood (with a lower pCO_2), where it combines with H_2O to produce carbonic acid, which immediately dissociates to bicarbonate and hydrogen ions. The relationship among pH, HCO_3^-, and pCO_2 (in mmHg) is shown in this equation:

$$pH = pK + \log[HCO_3^-/(0.03 \times pCO_2)] \qquad \text{Eqn (10)}$$

Note that the pK is defined as the pH at which the HCO_3^- and H_2CO_3 (or $0.03 \times pCO_2$) are in equal concentrations. That is, their ratio is 1, and the log of 1 is equal to 0.

At the normal concentration ratio in blood of 20:1 ("ideal" would be 1:1), and with a pK of 6.1 ("ideal" would be 7.4), the $HCO_3^- - H_2CO_3$ buffer system would seem to be ill-suited to buffering pH in blood. However, this excess of base (HCO_3^-), along with the volatility of CO_2, is tremendously able to prevent overaccumulation of acid. The lungs effectively make this an open system, with the loss of CO_2 providing almost unlimited buffering capacity. Bicarbonate is regulated primarily by the kidneys, and CO_2 is regulated by the lungs. It is the ratio of HCO_3^- to H_2CO_3 that determines the pH. Thus, $HCO_3^-:H_2CO_3$ in a concentration ratio of 15:0.75 has the same pH as a ratio of 20:1.0.

Hemoglobin

Hb acts as a buffer by transporting acid from the tissues to the lungs. A remarkable feature of Hb is that it increases its affinity for hydrogen ions (H^+) as it loses oxygen. That is, HHb, also called reduced Hb, has a greater affinity for H^+ than does O_2Hb. In tissue capillaries, O_2Hb enters an environment of low pO_2 and high acid. These conditions promote release of O_2, which also promotes binding of H^+. In the lungs, HHb encounters an environment of high pO_2 and low acid; this environment promotes the gain of O_2 and the release of H^+. These relationships are shown in Figure 1-1 *(12)*.

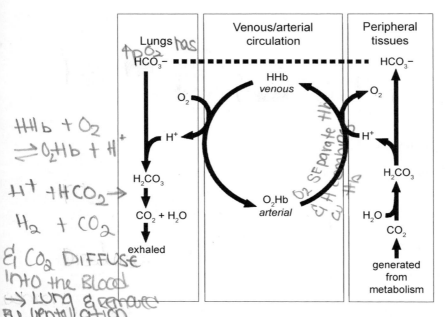

FIGURE 1-1. Interrelationships of the bicarbonate and hemoglobin buffering systems. *Reprinted with permission from Ehrmeyer SS, Shrout JB. Blood gases, pH, and buffer systems. In: Bishop ML, Duben-Engelkirk JL, Fody EP, eds. Clinical chemistry: principles, procedures, correlations, 2nd ed. Philadelphia: Lippincott, 1992.*

[handwritten at top: ALBUMIN, phosphate & protein ACT AS BUFFERS]

Phosphate

The HPO_4^{2-}-$H_2PO_4^-$ buffer pair is of minor importance as a buffer in plasma, with a concentration of ~1 mmol/L (3.1 mg/dL). It is of greater importance, and in higher concentration, as an intracellular buffer.

Albumin and other proteins

Largely due to the imidazole groups on the amino acid histidine, with a pK of ~7.4, albumin and other proteins can act as pH buffers. Because albumin is normally the major "unmeasured" anion in blood, albumin also affects the anion gap. For every 1 g/dL decrease in albumin below the patient's normal level (~4.4 g/dL), the AG will decrease by about 2.5 to 3 mmol/L. Patients in intensive care commonly have hypoalbuminemia, which can lower the AG and possibly confuse the interpretation of this parameter *(5,13)*.

[handwritten: FOR q ⊖ 1 g/dL IN ALBUMIN (~4.4 WNL) AG ↓ by 3 mmol/L]

ACID-BASE REGULATION

Respiratory (ventilatory) system

Because arterial CO_2 is influenced greatly by the ventilatory rate, pCO_2 is considered the respiratory component of the bicarbonate-CO_2 buffer system. Because CO_2 is the end product of many aerobic metabolic processes, buffering and removal of CO_2 are continually required for pH regulation. Provided there is a sufficient gradient of pCO_2 between tissues and blood, CO_2 will readily diffuse into the blood. CO_2 combines enzymatically with H_2O to form the unstable H_2CO_3, which quickly dissociates into HCO_3^- and H^+ ions. Recombination of HCO_3^- and H^+ would simply convert HCO_3^- back to CO_2. Instead, HHb plays a key role here by readily accepting the H^+ (see Figure 1-1).

As it enters a region of high pO_2 in the lungs, HHb picks up O_2 to become O_2Hb, which immediately promotes loss of H^+. H^+ quickly combines with HCO_3^- to produce dissolved CO_2, which diffuses into the alveolar air for ventilatory removal.

The arterial pCO_2 represents a balance between tissue production of CO_2 and pulmonary removal of CO_2. An elevated pCO_2 usually indicates inadequate ventilation (hypoventilation) and a respiratory acidosis. Conversely, a decreased pCO_2 usually indicates excessive ventilation (hyperventilation) and a respiratory alkalosis. There are several causes of respiratory abnormalities *(14)*.

Respiratory acidosis (ventilatory failure) can be caused by

- Obstructive lung disease (for example, chronic bronchitis or emphysema)
- Impaired function of respiratory center (head trauma, sedation, or anesthesia)
- Hypoventilation by mechanical ventilator

Respiratory alkalosis (hyperventilation) can be caused by

- Hypoxemia (which stimulates hyperventilation)
- Anxiety
- Hyperventilation by mechanical ventilator
- Metabolic acidosis
- Septicemia
- Trauma

Metabolic (renal) system

When the H^+ concentration deviates from normal, the kidneys respond by reabsorbing or secreting hydrogen, HCO_3^-, and other ions to regulate the pH of blood. Because kidneys are the major regulator of HCO_3^-, HCO_3^- is considered the metabolic component of the HCO_3^--CO_2 buffer system. (Although H^+ is regulated by the kidney, the liver, through gluconeogenesis, is several-fold as important as the kidney in removal of H^+.) Metabolic acidosis may develop if H^+ accumulates or if HCO_3^- ions are lost. Metabolic alkalosis may develop from either loss of H^+ or increase of HCO_3^-. Although a change in ventilatory rate can alter arterial pH in minutes, the kidneys require hours to days to significantly affect pH by altering the excretion of HCO_3^-. Thus, it takes several hours for these metabolic processes to alter the pH significantly.

Compensation

Compensation is a homeostatic response to an acid-base disorder in which the body attempts to restore pH to normal by adjusting the HCO_3^-: $(0.03 \times pCO_2)$ ratio back to a normal ratio of 20:1. Compensation involves either a relatively rapid ventilatory response (change in pCO_2) to a metabolic abnormality, or a relatively slow metabolic response (change in HCO_3^-) to a ventilatory abnormality. Here are some facts to remember about compensation:

- It is driven by changes in pH.
- As compensation returns pH to normal, the pH-driven compensation process slows, then stops.
- Significant respiratory compensation by loss of CO_2 occurs in 12–24 hours, while metabolic compensation by renal loss of HCO_3^- is much slower, taking 2–5 days for full compensation.
- Respiratory compensation by hyperventilation in metabolic acidosis is fairly predictable.
- Respiratory response to metabolic alkalosis is less predictable. While recent studies have shown that hypoventilation almost invariably occurs in metabolic alkalosis, other factors such as pain and hypoxemia can stimulate hyperventilation and overcome the hypoventilatory effect of metabolic alkalosis *(15)*.

Foundations of Blood Gases

The expected compensation for each acid-base abnormality will be discussed later in the sections "Clinical Abnormalities of Acid-Base Balance" and "Detecting Mixed Acid-Base Disorders."

HEMOGLOBIN BINDING

Hemoglobin binding to oxygen

Hb is a protein of 64,500 Da that consists of four heme molecules attached to four globin molecules. Hb is certainly a hall-of-fame molecule, having the essential ability both to transport and release oxygen to the tissues and to transport H^+ and carbon dioxide from the tissues to the lungs. Each of the four heme groups contains an Fe ion and can bind one molecule of oxygen. Structural changes occur with oxygen binding, and these changes result in color changes to the molecule. HHb gives blood a deep purplish hue, while O_2Hb makes blood appear scarlet red (14).

Exposing Hb to increasing concentrations of O_2 or increasing pO_2 will eventually cause Hb to become saturated with oxygen. The sigmoidal relationship between pO_2 and sO_2 is well known as the Hb-oxygen dissociation curve, shown in Figure 1-2.

Fetal Hb, a form of Hb present in decreasing amounts during the first month of life, persists in lower amounts for the first year of life and occasionally in

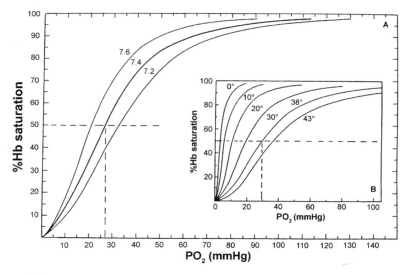

FIGURE 1-2. Oxyhemoglobin dissociation curve for whole blood. The curves on the left (A) are at a constant temperature (38 °C); the curves in the smaller box (B) are at a constant pH (7.4). To convert pO_2 in mmHg to kPa, multiply the value by 0.1333. *Reprinted with permission from Lambertson CJ. Transport of oxygen, carbon dioxide, and inert gases by the blood. In: Mountcastle VB. Medical physiology, 14th ed. St. Louis: CV Mosby, 1980:1725.*

thalassemias. Fetal Hb has a slightly altered absorbance spectrum, which can falsely elevate readings for COHb. Different wavelengths, as used on some new oximeters, and various correction factors have been proposed *(16)*, some of which are based on fully saturating blood with oxygen before measurement.

Although sO_2 in Equation 5 may be used for determining the adequacy of gas exchange in the lungs (as in pulse oximetry), pO_2 is better for this purpose. The %O_2Hb in Equation 6 is even less useful for monitoring pulmonary gas exchange, because a low %O_2Hb could be due to a high COHb, as in a heavy smoker or in cases of carbon monoxide poisoning.

Factors that affect hemoglobin affinity for oxygen

Several factors affect Hb binding of oxygen. In addition to pO_2, these factors are H^+, temperature, pCO_2, and 2,3-diphosphoglycerate (2,3-DPG). The small but amazing molecule nitric oxide (NO) also appears to play a role in oxygen delivery.

The effects of these factors are easy to remember because they are inversely related to oxygen binding. That is, an increase of H^+ concentration, temperature, carbon dioxide pressure, or 2,3-DPG concentration will decrease the affinity of Hb for oxygen. These factors also promote binding or release of oxygen in different areas of the body. For example, in the tissues, blood enters an area of relative warmth, acidity, elevated pCO_2, and low pO_2. These conditions promote loss of O_2 to tissues and binding of H^+ by Hb.

A major physiologic function of NO is regulating vascular tone, which can affect both blood perfusion and O_2 delivery to tissures. Structurally very similar to molecular oxygen (O_2), NO also binds to the heme iron of Hb and can form both methemoglobin and S-nitroso hemoglobin (SNO-Hb). This SNO-Hb may participate in delivering NO to tissues, especially those in hypoxic conditions. The binding and release of oxygen by Hb may selectively release NO to the cell membrane, where it affects blood pressure. The mechanisms of these reactions are complex *(17)*.

The substance 2,3-DPG has a separate function within the erythrocyte in altering Hb binding of oxygen by offsetting the effects of acidemia (2,3-DPG decreased) or alkalemia (2,3-DPG increased).

OXYGEN DELIVERY TO TISSUES

Gas exchange in the lungs

The pO_2 of atmospheric air (21% oxygen) is ~150 mmHg, accounting for the contribution of water vapor (45 mmHg) to the atmospheric pressure [(760 − 45) × 0.21 = 150 mmHg]. An adult breathing air with a pO_2 of 150 mmHg should have an alveolar pO_2 slightly above 100 mmHg and an arterial pO_2 of 80–100 mmHg. The range for arterial pO_2 depends on age *(14)*.

Inadequate delivery of oxygen to the lungs is usually the result of hypoventilation. Diminished gas exchange in the lungs may be caused by insufficient

Foundations of Blood Gases

surfactant (neonatal respiratory distress syndrome); pulmonary congestion, asthma, edema, inflammation, or fibrosis; mucus hypersecretion (bronchitis); or loss of alveolar compliance.

Blood flow to tissues

Although normal blood flow provides sufficient oxygen and nutrients to all organs and tissues, abnormalities in flow that cause inadequate perfusion of general or localized areas of the body may develop. Such conditions may lead to shock states, tissue necrosis, and ultimately death. Some common causes of poor perfusion:

- Inadequate cardiac output
- Decreased blood volume (hypovolemia)
- Emboli
- Vasoconstriction
- Shunting of blood

Both anemia and intrapulmonary shunting may diminish oxygen content in arterial blood. In anemia, blood with less Hb will carry proportionately less oxygen. In intrapulmonary shunting, venous blood passing through some pulmonary capillaries does not get oxygenated because these capillaries pass through nonfunctioning alveoli. When blood from these "shunts" mixes with oxygenated blood in the pulmonary vein, blood of a lower pO_2 and oxygen content is delivered to the systemic circulation *(14)*.

Release of oxygen to tissues

If arterial pO_2 and Hb content are normal and blood flow is adequate to a particular tissue, the conditions in the tissue of low pO_2 with high pCO_2 and acidity will usually ensure that oxygen is released and delivered to the tissues. The measurement of blood lactate may be used to determine whether overall oxygen metabolism is adequate.

CLINICAL ABNORMALITIES OF ACID-BASE BALANCE

Metabolic (Nonrespiratory) Acidosis

Metabolic acidosis is any clinical process that leads to a decreased blood pH (acidemia) and a decreased HCO_3^- level. It is caused by either a gain of a strong acid or loss of a base, usually bicarbonate.

Causes of metabolic acidosis *(18,19)*:

- Loss of bicarbonate: diarrhea, renal tubular acidosis

The following causes are usually associated with an increased anion gap:

- Excess production of acid from ketoacidosis, hypoxia, or lactate acidosis. If glucose cannot enter cells to produce adequate adenosine triphosphate (ATP), such as with insulin deficiency, fatty acids that produce ketoacids will be oxidized. If adequate oxygen is not available to tissues, cellular conditions begin to favor conversion of pyruvate to lactate, which produces far less ATP than by oxidative metabolism. Actually, lactate, not lactic acid, is produced *(11)*. The acid (H^+) comes from the degradation of large amounts of ATP.
- Ingestion of acids or acid-producing substances, for example, salicylates, ethanol, ethylene glycol, methanol, etc.
- Inadequate renal excretion of acids from normal metabolism, i.e., renal failure. Depending on whether glomerular or tubular failure occurs, accumulation of H^+ and/or loss of HCO_3^- may occur. Because renal acidosis usually has a slow onset, hyperventilation usually readily compensates to prevent acidemia.

Use of the anion gap in metabolic acidosis

The anion gap (AG), especially when elevated, may be useful in diagnosing the type of metabolic acidosis and in indicating the possibility of a mixed acid-base disorder, as shown in Table 1-3.

As described earlier, the AG is the difference between the commonly measured cations (Na and sometimes K) and the commonly measured anions (Cl and HCO_3^-). The approximate reference ranges for each are 8–16 mmol/L (8–16 mEq/L) for Na − (Cl + CO_2), and 12–20 mmol/L (12–20 mEq/L) for (Na + K) − (Cl + CO_2).

Expected compensation in metabolic acidosis (Table 1-5 and Figure 1-3)

In the initial phase of metabolic acidosis (acidemia with a low plasma bicarbonate), both pH and HCO_3^- are decreased and pCO_2 is still normal. The expected respiratory compensation is hyperventilation, which lowers blood pCO_2. This respiratory response can start within minutes, have a significant effect in 2 hours, and reach maximal effect in 12 to 24 hours. A general rule is that respiratory compensation lowers pCO_2 by 1.2 mmHg for each 1.0 mmol/L decrease in HCO_3^- below 24 mmol/L *(20)*. For example, in a patient who develops metabolic acidosis, after 24 hours, if the HCO_3^- is 16 mmol/L (decreased by 8 mmol/L), the pCO_2 should be approximately 30 mmHg (40 − 1.2 × 8). This is the equation for this relationship:

$$pCO_2 = 40 - 1.2 \times [24 - HCO_3^-] = 40 - 29 + 1.2 \times [HCO_3^-]$$

Simplified even further, the equation becomes

$$pCO_2 = 11 + 1.2 \times [HCO_3^-] \qquad \text{Eqn (11)}$$

Foundations of Blood Gases

TABLE 1-5. pH, HCO_3^-, and pCO_2 in Primary Disorders, Compensation, and Mixed Disorders

Primary disorder	pH	HCO_3^-	pCO_2	Condition
Metabolic acidosis	Dec	Dec	Norm	Early phase of metabolic acidosis (<2 h)
	Dec to norm	Dec	Dec	Expected respiratory compensation (≥6 h)
	Dec	Dec	Norm to dec	Inability to compensate
	Dec++	Dec	Inc	Mixed metabolic acidosis and respiratory acidosis
Metabolic alkalosis	Inc	Inc	Norm	Early phase of metabolic acidosis (<2 h)
	Inc to norm	Inc	Inc	Expected respiratory compensation (≥6 h)
	Inc	Inc	Norm to inc	Inability to compensate
	Inc++	Inc	Dec	Mixed metabolic alkalosis and respiratory alkalosis
Respiratory acidosis	Dec	Norm	Inc	Initial phase of respiratory acidosis
	Dec to norm	Inc	Inc	Expected metabolic compensation (≥24 h)
	Dec	Norm to inc	Inc	Inability to compensate
	Dec++	Dec	Inc	Mixed respiratory acidosis and metabolic acidosis
Respiratory alkalosis	Inc	Norm	Dec	Initial phase of respiratory alkalosis
	Inc to Norm	Dec	Dec	Expected metabolic compensation (≥24 h)
	Inc	Norm to dec	Dec	Inability to compensate
	Inc++	Inc	Dec	Mixed respiratory alkalosis and metabolic alkalosis

Dec, decreased; norm, normal; inc, increased; ++, markedly.

Thus, if the pCO_2 is >3 mmHg above this calculated value, the patient's lungs may not be fully capable of hyperventilating and may have an underlying respiratory acidosis. If the pCO_2 is <3 mmHg of this calculated value, an underlying respiratory alkalosis may be present. Either of these situations would indicate that a mixed disorder is present (metabolic acidosis with either respiratory acidosis or respiratory alkalosis).

Caution: These calculations for expected compensation should be used as guidelines in diagnosis and should not be over-interpreted as an absolute diagnosis.

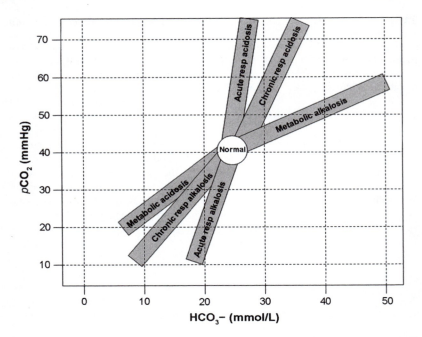

FIGURE 1-3. Expected relationships between HCO_3^- and pCO_2 during compensation for primary acid-base disorders. The areas shown are based on plots of equations 11, 12, 14, and 16.

Treatment

The best treatment for metabolic acidosis is correcting the underlying cause, such as by administering insulin, improving oxygen delivery, etc. If the condition must be corrected urgently, bicarbonate may be administered based on the blood bicarbonate level or the base deficit *(21)*.

Metabolic (Nonrespiratory) Alkalosis

Metabolic alkalosis is an acid-base disorder characterized by an elevation in bicarbonate level above 26 mmol/L and an elevation in pH. Metabolic alkalosis is frequently associated with renal impairment because a healthy kidney can excrete large amounts of HCO_3^- when it is in excess *(22)*.

Causes of metabolic alkalosis:

- **Loss of acid by vomiting or loss of acidic urine.** The initial loss of gastric acid causes metabolic alkalosis, which may enhance renal tubular loss of K^+ to conserve H^+.
- **Deficiency of Cl.** Deficiency of Cl enhances HCO_3^- reabsorption in the renal tubule. Because the reabsorption of cations (Na, K) requires an anion to follow, with less Cl available, more HCO_3^- is reabsorbed.

- **Hypokalemia.** Hypokalemia stimulates distal tubular reabsorption of HCO_3^- ions. In primary mineralocorticoid excess, aldosterone increases tubular Na reabsorption and promotes loss of K and H ions. These movements lead to extracellular alkalosis with hypochloremia, hypokalemia, and expanded ECF volume.
- **Administration of excess HCO_3^-.** Although not usually required for treatment of metabolic acidosis, infusion of sodium bicarbonate solution may sometimes be used. If excessive, HCO_3^- administration may lead to alkalosis, especially if renal function is compromised, as is often the case in cardiac failure.
- **Diuretics.** Loop diuretics interfere with tubular reabsorption of Na and Cl, which can lead to contraction of the extracellular space. Thiazide diuretics enhance K loss directly.
- **Corticosteroid excess.** Excess sodium-retaining steroids, such as cortisol, lead to excretion of both K^+ and H^+ in the distal renal tubules. Excess cortisol, as in Cushing's syndrome, can lead to both metabolic alkalosis and hypokalemia, because cortisol and other steroids such as corticosterone have some mineralocorticoid activity.
- According to the SID concept *(7)*, a decrease in plasma albumin (being the predominant weak acid in plasma) causes a decreased weak acid concentration that results in a metabolic alkalosis.

Expected compensation in metabolic alkalosis (Table 1-5 and Figure 1-3)

As stated earlier, whereas metabolic alkalosis consistently causes hypoventilation, the actual compensation response for metabolic alkalosis is not always predictable. Although recent studies have shown that hypoventilation almost invariably occurs in metabolic alkalosis, other factors such as pain and hypoxemia can stimulate hyperventilation and overcome the hypoventilatory stimulus of metabolic alkalosis *(15)*. Since alkalosis also increases the affinity of Hb for O_2 (left shift), oxygen is less available to the tissues and the compensatory hypoventilation can further decrease oxygen availability and is a clinical consideration. Although the maximal respiratory compensation by hypoventilation was once believed to be no more than 55–60 mmHg, compensation can elevate arterial pCO_2 to over 80 mmHg *(15)*.

In the initial phase of metabolic alkalosis, both pH and HCO_3^- are increased and pCO_2 remains normal. Within 2 hours, compensation by hypoventilation is expected, but maximal compensation requires 12 to 24 hours, which increases the blood pCO_2 to match the elevated HCO_3^- and restores the HCO_3^-/pCO_2 ratio to normal. In general, respiratory compensation increases pCO_2 by 0.7 mmHg for each 1.0 mmol/L rise in HCO_3^-. For example, in a patient who develops metabolic alkalosis, after 24 hours, if the HCO_3^- is 32 mmol/L (8 mmol/L above a very normal 24 mmol/L), the pCO_2 should be approximately 46 mmHg (40 + 0.7 × 8) *(23)*. Here is the relevant equation:

$$pCO_2 = 40 + 0.7 \times [HCO_3^- - 24] = 40 + 0.7 \times HCO_3^- - 17$$

This simplifies further to

$$pCO_2 = 23 + 0.7 \times HCO_3^- \qquad \text{Eqn (12)}$$

If the actual pCO_2 deviates from this expected pCO_2 by more than 4 mmHg, the patient may have a mixed disorder due to either an underlying respiratory acidosis (if more than 4 mmHg higher) or alkalosis (if more than 4 mmHg lower).

Caution: These calculations for expected compensation should be used as guidelines in diagnosis and should not be over-interpreted as an absolute diagnosis.

Treatment

The optimal treatment for metabolic alkalosis is correcting the underlying cause. In some cases, simple hydration will gradually correct metabolic alkalosis if renal function is normal. If Cl depletion is present, Cl must be replaced by administration of NaCl or KCl as appropriate, with consideration of the renal function. Rarely, diluted HCl may be administered intravenously if necessary *(24)*.

Respiratory Acidosis

A decreased pH and increased pCO_2 are diagnostic for respiratory acidosis (ventilatory failure) such that the pCO_2 in blood rises to abnormal levels and pH falls below 7.35. Acute respiratory acidosis occurs by one of these processes: decreased breathing or ventilation, increased production of CO_2 by the body, or excess CO_2 in the inspired gas *(25)*. Normally, any increased production of CO_2 promptly stimulates hyperventilation, which maintains arterial pCO_2 at normal levels. If arterial pCO_2 is increased, it almost always involves a problem with ventilation.

Causes of respiratory acidosis:

- Impaired pulmonary gas exchange, such as chronic bronchitis, emphysema, or asthma. Respiratory acidosis is the end result of chronic progressive alveolar destruction.
- Acute airway obstruction caused by aspiration or blockage of a tube.
- Circulatory failure, which causes insufficient delivery of blood to the lungs.
- Impaired function of respiratory center caused by head trauma, drugs, toxins, or anesthesia.
- Mechanical ventilation that does not provide adequate oxygen for the patient. If CO_2 production increases while ventilation rate remains constant, the blood pCO_2 will rise *(26)*.

Expected compensation in respiratory acidosis (Table 1-5 and Figure 1-3)

During the acute phase, plasma buffering of the elevated CO_2 increases the HCO_3^- slightly, by approximately 1 mmol/L for each 10 mmHg rise in pCO_2. Over the next few hours, the kidneys increase reabsorption of HCO_3^-, which elevates serum HCO_3^- by about 2 mmol/L. As this continues into the chronic

Foundations of Blood Gases

phase (over 24 hours), the HCO_3^- slowly rises and plateaus after 2–5 days with normalization of the pH. Since compensation occurs in two phases, the expected compensation has two algorithms to predict the expected blood level of HCO_3^- *(27)*:

1. Acute Resp Acid: For each 10 mmHg rise in arterial pCO_2 above 40 mmHg, HCO_3^- should increase by about 1 mmol/L and pH should decrease by about 0.07–0.08. For example, if a patient's breathing slows for a few hours and the pCO_2 is 50 mmHg, the HCO_3^- should rise by only 1 mmol/L (i.e., 25 mmol/L) and the pH should decrease by 0.07 (pH of about 7.33).

$$\text{pH change} = 0.007 \times (40 - pCO_2) \qquad \text{Eqn (13)}$$

$$HCO_3^- = 24 + 0.1 \times (pCO_2 - 40), \text{ which rearranges to}$$

$$HCO_3^- = 20 + 0.1 \times pCO_2 \qquad \text{Eqn (14)}$$

2. Chronic Resp Acid: The maximal HCO_3^- response occurs in about 3 days, with the HCO_3^- increasing by 3-4 mmol/L for each 10 mmHg rise in arterial pCO_2. For example, if the pCO_2 has increased from 40 mmHg to a steady state of 60 mmHg for several days, the HCO_3^- should have risen about 8 mmol/L. That is, if the patient's normal HCO_3^- was 24 mmol/L, it would be about 32 mmol/L after 3 days of metabolic compensation. This will also normalize the pH, so that pH and bicarbonate may be calculated in this way:

$$\text{pH change} = 0.003 \times (40 - pCO_2) \qquad \text{Eqn (15)}$$

$$HCO_3^- = 24 + 0.35 \times (pCO_2 - 40), \text{ which rearranges to:}$$

$$HCO_3^- = 10 + 0.35 \times pCO_2 \qquad \text{Eqn (16)}$$

Note that these calculations are time dependent and assume the patient started with a very normal pH (7.40), pCO_2 (40 mmHg), and HCO_3^- (24 mmol/L).

Caution: These calculations for expected compensation should be used as guidelines in diagnosis and should not be over-interpreted as an absolute diagnosis.

Treatment

Adequate lung ventilation must be restored by such means as endotracheal intubation or positive pressure ventilation. Care must be taken to avoid too rapid correction of the elevated pCO_2. Administration of bicarbonate is rarely done, because HCO_3^- crosses the blood-brain barrier slowly, which can elevate the blood pH without affecting the CNS pH *(28)*.

Respiratory Alkalosis

Respiratory alkalosis is associated with a fall in blood pCO_2 to below 35 mmHg and an elevated blood pH. Respiratory alkalosis can be caused by any condition that leads to hyperventilation. From a normal ventilatory rate of 12–15/min, respiratory alkalosis (hyperventilation) results from an increased ventilatory rate ($\geq$20/min) in which CO_2 is lost faster than it is produced, leading to an increased pH and decreased pCO_2.

Causes of respiratory alkalosis:

- Hypoxemia-induced hyperventilation, as caused by breathing oxygen-poor air or exposure to high altitude.
- Pulmonary embolism or pulmonary edema, in which oxygen transport across the alveolar membrane is impaired to a greater extent than is CO_2 transport.
- Anxiety-induced hyperventilation.
- Excess mechanical ventilation, usually from aggressive use of the ventilator to increase arterial oxygen tension.
- Drugs such as salicylate, nicotine, or progesterone, which can cause hyperventilation.
- Central nervous system disorders that result from conditions such as sepsis or trauma.

Expected compensation in respiratory alkalosis (Table 1-5 and Figure 1-3)

The expected metabolic response to the decreased pCO_2 in respiratory alkalosis is for the kidney to increase excretion of HCO_3^-. Since metabolic compensation is relatively slow, it is time-dependent and occurs in two phases. Thus, the expected compensation has two algorithms to predict the expected blood level of HCO_3^- *(29)*:

1. Acute Resp Alkalosis (<24 h): For each 10 mmHg fall in arterial pCO_2 below 40 mmHg, the pH should increase by about 0.08, and HCO_3^- should decrease by ~2 mmol/L. As an example, during this phase, if pCO_2 is 30 mmHg, this 10 mmHg fall should be associated with a pH increase of 0.08 (pH ~ 7.48) and a bicarbonate decrease of 2 mmol/L (HCO_3^- ~ 22 mmol/L).

$$HCO_3^- = 24 - 0.2 \times (40 - pCO_2), \text{ which simplifies to}$$
$$HCO_3^- = 16 + 0.2 \times pCO_2 \qquad \text{Eqn (17)}$$

2. Chronic Resp Alkalosis (2–5 days): pH should be increased by about 0.03 for each 10 mmHg fall in pCO_2 and HCO_3^- should be decreased by about

5 mmol/L. For example, if the pCO_2 is 30 mmHg after 2 days of hyperventilation, the pH should be increased by 0.03 (pH ~7.43) and HCO_3^- should be decreased by about 5 (19 mmol/L).

$$HCO_3^- = 24 - 0.5 \times (40 - pCO_2)$$, which simplifies to

$$HCO_3^- = 4 + 0.5 \times pCO_2 \qquad \text{Eqn (18)}$$

Note: The lower limit for metabolic compensation in hyperventilation is a plasma HCO_3^- of ~12 mmol/L (~12 mEq/L). If plasma HCO_3^- is <12 mmol/L (<12 mEq/L), an underlying metabolic acidosis may be present.

Caution: These calculations for expected compensation should be used as guidelines in diagnosis and should not be over-interpreted as an absolute diagnosis.

Treatment

Treatment is aimed at correcting the underlying condition. Most urgently, hypoxemia must be corrected by giving oxygen. Salicylate overdose should be treated. Any anxiety-induced respiratory alkalosis is best treated with reassurance. Sometimes, rebreathing expired air (higher CO_2 content) can be helpful in otherwise healthy persons *(30)*.

Detecting Mixed Acid-Base Disorders

Mixed acid-base disorders occur when multiple primary acid-base disorders occur at the same time. These are common in hospital populations, especially in the ED and in critical care. Although there are equations and diagrams that can aid in the diagnosis, these relationships will not hold in patients with complex and long-standing acid-base disorders. There is no substitute for a careful review of a patient's clinical course.

The importance of identifying mixed acid-base disorders lies in their diagnostic and therapeutic implications. For example, the development of a primary metabolic alkalosis in a patient with chronic obstructive airway disease who is being treated with diuretics should alert the clinician to possible potassium depletion, and a patient presenting with a mixed respiratory alkalosis and metabolic acidosis should be evaluated for salicylate intoxication *(31)*.

Does the expected compensation occur?

A simple concept to remember in a primary acid-base disorder is that, if the expected compensation does not occur, a mixed disorder should be suspected. For example, in a primary metabolic acidosis, the lungs are expected to compensate this excess metabolic acid by hyperventilating (a respiratory alkalotic process) to remove the respiratory acid CO_2 and return the pH towards normal. If this does not occur as expected, the person is considered to have an underlying respiratory acidosis *(31)*.

Metabolic acidosis

The body's natural response to compensate metabolic acidosis (acidemia with a low plasma bicarbonate) during the first 12 to 24 hours is hyperventilation, which decreases the blood pCO_2 by 1.2 mmHg for each 1 mmol/L drop in HCO_3^-. As derived earlier *(20)*, the expected pCO_2 (mmHg) during this time should be within 3 mmHg of

$$pCO_2 = 11 + 1.2 \times [HCO_3^-] \qquad \text{Eqn (11)}$$

Metabolic alkalosis

The respiratory compensation to metabolic alkalosis is hypoventilation, which increases the pCO_2. As described earlier *(23)*, during this period of compensation, the expected pCO_2 (mmHg) should be within 3 mmHg of

$$pCO_2 = 23 + 0.7 \times HCO_3^- \qquad \text{Eqn (12)}$$

Figure 1-3 shows the expected areas of pCO_2 vs HCO_3^- for both metabolic acidosis and metabolic alkalosis. However, as mentioned previously, expected compensation calculated by an equation should only be used as a guide and not as an absolute diagnosis. This is especially true when a patient has had multiple (mixed) acid-base disorders that have occurred over a period of time.

Respiratory acidosis or respiratory alkalosis

If a patient has only a primary respiratory disorder, either acidosis or alkalosis, simple rules predict the expected pH changes vs. the change in pCO_2 during the acute and the chronic phases. During the acute phase of respiratory disorders, for each 10 mmHg rise or fall in pCO_2, the pH should change by 0.08. As this progresses into the chronic phase (1–2 days), for each 10 mmHg rise or fall in pCO_2, the pH should have changed by only 0.03 *(29)*. Table 1-6 illustrates the relationships between pCO_2, pH, and HCO_3^- for acute and chronic respiratory changes.

TABLE 1-6. Expected Changes for pCO_2, pH, and HCO_3^- for Acute and Chronic Respiratory Conditions

pCO_2 (mmHg)	Expected changes in acute respiratory conditions		Expected changes in chronic respiratory conditions	
	pH	HCO_3^- (mmol/L)	pH	HCO_3^- (mmol/L)
70	7.16	27	7.31	34.5
60	7.24	26	7.34	31
50	7.32	25	7.37	27.5
40[a]	**7.40**[a]	**24**[a]	**7.40**[a]	**24**[a]
30	7.48	22	7.43	19
20	7.56	20	7.46	14

[a]Bolded values indicate normal values for each parameter.

Foundations of Blood Gases

Figure 1-3 is a graphical plot of the equations for predicting how bicarbonate is expected to change during acute and chronic primary respiratory disorders and how pCO_2 is expected to change during primary metabolic disorders with compensation. If the blood gas data clearly do not reflect these temporal relationships, it suggests that an additional disorder is present and that the patient has a mixed disorder. However, these relationships are guidelines only and may not hold for complex acid-base disorders.

Delta ratio

The "delta ratio" is the ratio of the increase in AG (Na − Cl − HCO_3^-) from a mid-normal value such as 12, to the decrease in HCO_3^- from a normal value such as 24 *(32)*. Here is an equation for delta ratio:

$$\text{Delta Ratio} = (AG - 12) / (24 - HCO_3^-) \qquad \text{Eqn (19)}$$

In a typical metabolic acidosis, the increase in AG from 12 mmol/L and the decrease in HCO_3^- from 24 mmol/L should equal each other, so the "delta ratio" would be one. If the delta ratio is greater than one, it suggests there is more HCO_3^- in the blood than expected from the increased AG, so that an additional metabolic alkalosis may be present. If the delta ratio is less than one, it suggests less HCO_3^- is present than expected from the increased AG, so that an additional non-AG metabolic acidosis is present. This is illustrated in Table 1-7.

Here are some other tips for diagnosing mixed acid-base disorders:

- When the pH is well within normal and both the HCO_3^- and pCO_2 are abnormal. This strongly suggests a mixed disorder.
- Whenever HCO_3^- and pCO_2 are abnormal in the opposite direction. The compensatory response should always be in the same direction as the change caused by the primary disorder.
- Simple acid-base disorders do not overcompensate. That is, compensation will not cause an acidemia to become an alkalemia.

TABLE 1-7. Examples of Delta Ratio in Metabolic Acidoses

Anion gap (mmol/L)	HCO_3^- (mmol/L)	Delta ratio	Condition(s)
20	16	1.0	Pure AG metabolic acidosis
20	21	2.7	Mixed disorder: AG metabolic acidosis + metabolic alkalosis
20	13	0.7	Mixed disorder: AG metabolic acidosis + non-AG metabolic acidosis

Disorders of Oxygenation: Hypoxemia and Tissue Hypoxia

The section "Evaluation of Oxygenation Status" will discuss low-oxygen states in detail. Inadequate oxygenation of blood may be caused by

- Insufficient oxygen in alveolar air
- Pulmonary embolism
- Impaired gas exchange in the lungs between alveolar air and blood

Poor tissue oxygenation may be caused by

- Inadequate blood flow to tissues because of diminished cardiac output or altered blood flow
- Inadequate uptake of oxygen by the tissues

Both pO_2 and blood lactate may be used to evaluate these conditions.

EVALUATION OF BLOOD GAS RESULTS

Reference Ranges

The reference ranges for both arterial and venous blood are shown in Table 1-1 for the common blood gas and acid-base parameters *(9,14)*. The clinically acceptable range for arterial blood is also shown, which may be used as a guideline in critical care *(14)*. Note that the pO_2 of arterial blood varies with age (see Table 1-8).

Evaluating Acid-Base Status

Ideally, evaluating acid-base disorders, including mixed acid-base disorders, would include simultaneous information on the patient's clinical history, the electrolyte results, and the blood gas results. However, in practice, the blood gas results are usually available before the electrolyte results *(33)*.

The acid-base and blood gas results (pH, pCO_2, pO_2, HCO_3^-) may be evaluated in several steps, as follows:

1. Evaluate the patient's clinical history and current status to anticipate conditions associated with acid-base disorders.
2. Evaluate the pH.
3. Evaluate the ventilatory (pCO_2) and metabolic (HCO_3^-) status to determine the primary disorder.
4. Evaluate laboratory and clinical data for a possible mixed disorder:
   a. For a primary metabolic (or respiratory) disturbance, is respiratory (or metabolic) compensation appropriate?
   b. Do any other parameters—such as electrolytes, lactate, AG, or delta ratio—indicate that another acid-base disturbance is present?
   c. Is the patient's history consistent with the blood gas results?

The evaluation of oxygen status is mostly independent of the acid-base interpretation, so this concept will be presented in a separate section.

Foundations of Blood Gases

TABLE 1-8. Acceptable Arterial Oxygen Tensions by Age

Age (y)	Acceptable $pO2$
1–60	>80 mmHg
70	>70 mmHg
80	>60 mmHg
90	>50 mmHg

Step 1. Evaluating the patient's status and history to anticipate possible acid-base abnormalities

There are many conditions associated with acid-base disorders, as listed in Table 1-9. A perceptive clinical evaluation can be the most important part of evaluating a patient for acid-base disorders.

Step 2. Evaluating the pH

An abnormal pH indicates that an acidosis or alkalosis has occurred and the extent of the acid-base disorder, and it may suggest that compensation has occurred. However, the pH by itself does not indicate whether a mixed disorder is present. Consider the following examples:

TABLE 1-9. Examples of the Expected Acid-Base Disorder Associated with Various Clinical Conditions

Clinical condition	Expected acid-base disorder
Cardiac arrest	Metabolic acidosis
Pulmonary arrest	Respiratory acidosis
Hyperventilation (many causes)	Respiratory alkalosis
Blockage of endotrachial tube	Respiratory acidosis
(after removal of blockage from tube)	(Respiratory alkalosis)
Congestive heart failure leading to hyperventilation	Respiratory alkalosis
Vomiting	Metabolic alkalosis
Diarrhea	Metabolic acidosis
Shock state (inadequate perfusion)	Metabolic acidosis
Pulmonary edema	Respiratory alkalosis (hypoxia leads to hyperventilation)
Severe pulmonary edema	Respiratory acidosis
Diuretic therapy	Metabolic alkalosis
Drug intoxication	Respiratory acidosis (respiratory arrest)
Bicarbonate therapy	Metabolic alkalosis
Poor perfusion	Metabolic acidosis

- pH 7.20 confirms that a severe acidosis is present and that compensation is either in its early stages or is ineffective in controlling the acidosis. Further investigation is required to determine the metabolic or respiratory origin of the acidosis and whether more than one acidotic process is present.
- pH 7.48 indicates that a mild alkalosis is present. Further information is required to determine the cause of the alkalosis and whether the alkalosis is in its early stages (and may get worse), has been nearly compensated, or is part of a mixed disorder.

A normal pH may indicate that the patient has no acid-base disorder. However, the patient may have a mix of acidotic and alkalotic events (primary or compensatory) that have offset each other. In addition to the patient's history, the HCO_3^- and pCO_2 must be considered, as in these examples:

- pH 7.45 with elevated HCO_3^- and pCO_2 suggests that a primary metabolic alkalosis has been compensated by an appropriate respiratory acidotic response. Similarly, a pH of 7.45 with decreased HCO_3^- and pCO_2 suggests that a primary respiratory alkalosis has been compensated by a metabolic acidotic response.
- pH 7.40 with both HCO_3^- and pCO_2 abnormal indicates that a mixed acidosis and alkalosis are present that are fortuitously offsetting each other to give normal pH.

Clinical information on the time course of these processes may confirm the more likely condition.

Step 3. Evaluate the ventilatory and metabolic status

The simplest approach to evaluating the pCO_2 and HCO_3^- is to consider each parameter separately as indicating acidosis, alkalosis, or normal status, then evaluate them along with pH to determine whether the primary disorder is respiratory or metabolic and whether a mixed disorder or compensation is present.

A decreased pCO_2 indicates a respiratory alkalotic process, which may be either primary or compensatory. An increased pCO_2 indicates a respiratory acidotic process, either primary or compensatory (Table 1-10).

A decreased HCO_3^- indicates a metabolic acidosis, either primary or compensatory. An increased HCO_3^- indicates a metabolic alkalosis, either primary or compensatory (Table 1-11).

Note: Used correctly, the terms "respiratory alkalosis," "metabolic acidosis," etc., refer to pathologic processes and not to compensation. However,

TABLE 1-10. Evaluation of Ventilatory Status by Arterial pCO_2

pCO_2	Ventilatory status
Decreased	Respiratory alkalotic process (hyperventilation) is present
Normal	Normal ventilatory status? (Is a mixed disorder present?)
Increased	Respiratory acidotic process (hypoventilation) is present

Foundations of Blood Gases

TABLE 1-11. Evaluation of Metabolic Status by Plasma Bicarbonate (HCO_3^-)

Bicarbonate (total CO_2)	Metabolic status
Decreased	A metabolic acidotic process is present (either primary, compensatory, or mixed)
Normal	No metabolic acid-base disorder present? (mixed disorder also possible)
Increased	A metabolic alkalotic process is present (either primary, compensatory, or mixed)

I believe that it simplifies interpretation to momentarily regard any decreased pCO_2 (for example) as a respiratory alkalotic process. Whether it is a primary abnormality or appropriate compensation will become clear as other factors are considered.

A normal pCO_2 and HCO_3^- indicate normal ventilatory and metabolic status. However, if one of these parameters is normal when the other is abnormal, then a mixed disorder may be present. That is, a lack of an appropriate compensatory response suggests that an additional disorder is present.

Consider an example

A pCO_2 result of 25 mmHg indicates that a respiratory alkalotic process is present. Whether it is a primary abnormality or appropriate compensation depends on the HCO_3^- and pH associated with this pCO_2. If the HCO_3^- is 14 mmol/L (14 mEq/L) (with a pH of ~7.32), then the pCO_2 of 25 mmHg probably indicates compensating hyperventilation (a respiratory alkalotic process) for a primary metabolic acidosis. In another example, if the HCO_3^- is 23 mmol/L (23 mEq/L) and the pH is ~7.53, the pCO_2 of 25 mmHg indicates a primary respiratory alkalosis, which could be from either acute or chronic hyperventilation, depending on the duration of the hyperventilation. If acute, the metabolic compensatory response to eliminate HCO_3^- is progressing about as expected. If the hyperventilation is chronic, then the higher than expected HCO_3^- suggests that, because the kidney is unable to adequately compensate by eliminating HCO_3^-, a metabolic alkalosis is also present.

Step 4. Evaluate for a possible mixed disorder

Step 4 has three components.

4a. Is the compensation adequate for the primary disorder? To evaluate whether or not the expected compensation is occurring for the primary acid-base disorder, refer to the "Expected Compensation" descriptions in the various sections of acid-base disorders described in "Clinical Abnormalities of Acid-Base Balance," and the section "Detecting Mixed Acid-Base Disorders."

Figure 1-4 is presented as a nomogram for understanding and interpreting the changes in acid-base results that occur in the various processes shown. The areas shown in Figure 1-4 have been derived from equations

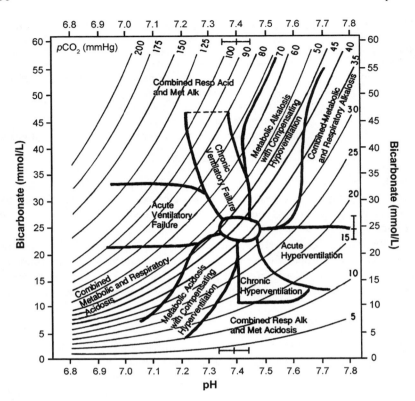

FIGURE 1-4. Nomogram for interpreting acid-base status. Curved lines represent a constant pCO_2. Met Alk, metabolic alkalosis; Resp Alk, respiratory alkalosis; Resp Acid, respiratory acidosis.

listed here and reported elsewhere *(33)*. Although this figure may be useful as a guide, actual diagnosis of a patient's condition must always include a clinical assessment of the patient's history and physical findings at the time the result was obtained. This nomogram is reasonably helpful in simple or mixed double disorders that occur simultaneously. It is important to note that Figure 1-4 cannot be used in any situation in which a new disorder develops in addition to existing disorders.

4b. **Do other laboratory results suggest that an additional acid-base disorder is present?** Several common laboratory tests may help determine whether additional acid-base disorders are present *(26,33)*.

- **Potassium.** Hypokalemia suggests a metabolic alkalosis may be present.
- **pH.** A normal pH combined with abnormal results for both HCO_3^- and pCO_2 warrants consideration of a mixed acid-base disorder.
- **Chloride.** Hyperchloremia results from HCO_3^- loss, either from a primary metabolic acidosis, or from the metabolic acidotic compensation for respiratory alkalosis. Hypochloremia may cause metabolic alkalosis

Foundations of Blood Gases

by promoting renal reabsorption of HCO_3^-, or may be caused by compensation for chronic respiratory acidosis (as HCO_3^- is retained and Cl^- is excreted).

- **Anion gap.** An elevated AG [(Na) − (Cl + HCO_3^-)] of >20 mmol/L (>20 mEq/L) indicates metabolic acidosis. AG is also useful in differentiating the causes of metabolic acidoses, as shown in Table 1-3 *(5)*.

    Elevated AG occurs in uremic acidosis, ketoacidosis, lactate acidosis, and ingestion of toxin (for example, glycol or salicylate). Normal AG occurs in renal tubular acidosis and diarrhea, among other conditions.

    If a metabolic acidosis is present, the presence or absence of an elevated AG will help in the differential diagnosis of metabolic acidosis.

    While often very useful, AG is calculated from three or four measurements and is subject to variation from these measurements. One report claimed that about half of patients with a clearly elevated blood lactate did not have an elevated AG *(6)*.

- **Delta ratio.** The "delta ratio" should also be evaluated, as discussed earlier in the section "Detecting Mixed Acid-Base Disorders." The "delta ratio" is the ratio of the increase in AG to the decrease in HCO_3^-. If the delta ratio is greater than one, it suggests there is more HCO_3^- in the blood (i.e., less is lost) than expected from the increased AG, so that an additional metabolic alkalosis may be present. If the delta ratio is less than one, it suggests less HCO_3^- is present (more is lost) than expected from the increased AG, so that an additional non-AG metabolic acidosis may be present.

- **Lactate.** A rising blood lactate is a sensitive indicator of metabolic acidosis resulting from poor oxygenation of tissues.

- **Creatinine.** An elevated creatinine indicates renal insufficiency and possible uremic metabolic acidosis.

4c. Does the patient have other conditions associated with an acid-base disorder? An evaluation of blood gas results must always include the patient's historical and physical findings. Many conditions are frequently associated with an acid-base disorder, with some common ones listed in Table 1-9, especially if the blood gas results cannot readily be explained by the patient's "primary" disorder. The presence of more than one such condition should alert the clinician to the possibility of a mixed acid-base disorder. Mixed acid-base conditions are quite common.

EVALUATION OF OXYGENATION STATUS

Evaluating Arterial Oxygenation

Three primary and four additional blood gas parameters are often used for assessing arterial oxygenation:

1. The arterial oxygen tension (pO_2), especially in relation to the oxygen content of the air the patient is breathing (FIO_2).

2. The percent oxygen saturation (sO_2) or the percent oxyhemoglobin (%O_2Hb) in blood
3. The hemoglobin concentration in blood

Some additional parameters may be calculated from the measured blood gas results:

4. The oxygen content of blood
5. Alveolar – arterial pO_2 difference or gradient
6. Intrapulmonary shunting of blood, in which blood flows into the lung, but does not perfuse functional alveoli
7. Tissue oxygenation

1 and 2. Arterial oxygen tension and oxygen saturation of hemoglobin

The pO_2 and sO_2 values are most often used for assessing arterial oxygenation. Each provides useful clinical information, with the pO_2 being a more linear reflection of the efficiency of blood oxygenation by the lungs, and the sO_2 being a better indicator of the O_2 content of blood. Hypoxemia is defined as an arterial pO_2 that is below the minimally acceptable limit for persons of a given age group breathing room air, as shown in Table 1-8 *(14)*. As pO_2 decreases to <60 mmHg, the degree of hypoxemia intensifies, with severe hypoxemia occurring below 40 mmHg (Table 1-12).

Note that the ideal diagnosis of hypoxemia should be made when the patient is breathing room air. However, if the patient is receiving oxygen therapy, it must not be interrupted simply to assess hypoxemia *(14)*. Instead, the likelihood of hypoxemia in a patient breathing oxygen-enriched air may be predicted by comparing the arterial pO_2 to the FI-O_2 (percent oxygen being breathed). If the arterial pO_2 (mmHg) is less than five times the FIO_2 (%), the patient's alveolar ventilation is likely inadequate and the patient would probably be hypoxemic if breathing room air (see Table 1-13).

TABLE 1-12. Evaluation of Hypoxemia in Patients <60 y Old (Breathing Room Air)

Arterial pO_2[a]	Condition
>80 mmHg	Adequate oxygenation
60–80 mmHg	Mild hypoxemia
40–60 mmHg	Moderate hypoxemia
<40 mmHg	Severe hypoxemia

[a]For each year over 60 y, subtract 1 mmHg from the limits for mild and moderate hypoxemia. A $pO2$ <40 mmHg at any age indicates severe hypoxemia. *Adapted with permission from Shapiro BA, Harrison RA, Cane RD, Templin R. Clinical applications of blood gases, 4th ed. St. Louis: CV Mosby, 1989.*

Foundations of Blood Gases

TABLE 1-13. Effect of Breathing Oxygen-Enriched Air

FI-O_2 (%)	Predicted minimal pO_2
30	150 mmHg
40	200 mmHg
50	250 mmHg
80	400 mmHg
100	500 mmHg

Adapted with permission from Shapiro BA, Harrison RA, Cane RD, Templin R. Clinical applications of blood gases, 4th ed. St. Louis: CV Mosby, 1989.

3. Hemoglobin concentration

Because a normal hemoglobin concentration in blood (12–17 g/dL depending on gender) represents an excess oxygen-carrying capacity, a slightly decreased Hb of 10 g/dL is usually of no concern in critical care. Clinicians usually want to be sure the person is not severely anemic.

4. Blood O_2 content

Blood oxygen content is calculated from measurements of blood Hb and sO_2. Anemia will dramatically affect O_2 content, as will hypoxemia (low $pO_2/sO2$), hypercapnia (high pCO_2), acidemia (low pH), and hyperthermia, all of which cause Hb to be less saturated for a given pO_2.

5. Alveolar-arterial oxygen gradient (A − a)

The alveolar-arterial oxygen gradient (A − a) may be useful for determining if there is ventilatory failure or if there is oxygenation failure *(34)*. The alveolar pO_2 (p_AO_2) may be calculated as the difference between the atmospheric pO_2 of 150 mmHg [(760 − 45) × 0.21 = 150 mmHg] (45 mmHg is atmospheric water vapor pressure) and the blood pCO_2 divided by a gas exchange factor (0.8). Thus, for typical conditions in a healthy person with a pCO_2 of 40 mmHg and a measured arterial pO_2 of 95 mmHg (paO_2), the alveolar-arterial pO_2 difference is this:

$$p_AO_2 - paO_2 = (150 - 40/0.8) - 95 = 5 \text{ mmHg}$$

As a guide to showing the usefulness of the A − a gradient in determining if ventilatory failure or oxygenation failure is present (if ventilation and perfusion are in harmony), see Table 1-14.

6. Determining intrapulmonary shunting

Intrapulmonary shunting of blood occurs when some portion of blood entering the lungs bypasses functional alveoli. Normally, about 4% of blood entering the lung passes into smaller vessels in the lung that bypass the

TABLE 1-14. Use of Alveolar-Arterial Difference (A – a) in Evaluating Oxygenation and Ventilation Failure

Parameter	Normal oxygenation and ventilation	Oxygenation failure	Ventilation failure
paO_2 (mmHg)	95	45	45
pCO_2 (mmHg)	40	40	80
A – a gradient (mmHg)	5 (normal)	55 (elevated)	5 (normal)

alveoli. If collapsed, nonfunctional, or nonventilated alveoli are present, this percentage will increase. The percent of shunted blood may be calculated as the ratio of shunted cardiac output to total cardiac output *(14)*:

$$\frac{O_2 \text{ content of pulmonary capillary blood} - O_2 \text{ content of arterial blood}}{O_2 \text{ content of pulmonary capillary blood} - O_2 \text{ content of venous blood}}$$

This equation may be simplified to using only the oxygen saturation (sO_2) values of the blood locations:

$$\frac{sO_2 \text{ of pulmonary capillary blood} - sO_2 \text{ of arterial blood}}{sO_2 \text{ of pulmonary capillary blood} - sO_2 \text{ of venous blood}}$$

The O_2 content of pulmonary capillary blood should be the highest of any blood in the circulation, so the sO_2 is assumed to be 100%. With sO_2 measured on both arterial and venous blood, the equation becomes

$$\frac{100 - sO_2 \text{ of arterial blood}}{100 - sO_2 \text{ of venous blood}}$$

The relationship between arterial and venous sO_2 and Intrapulmonary Shunt Fraction is illustrated in Table 1-15.

7. Evaluating tissue oxygenation

Tissue oxygenation can be assessed from calculations of O_2 delivery (DO_2) and O_2 consumption (VO_2). In addition to measurement of blood oxygen content, both DO_2 and VO_2 require measurement of cardiac output. Although cardiac output may be crudely assessed by measurements such as blood pressure,

TABLE 1-15. Use of Arterial and Venous sO_2 in Evaluating Intrapulmonary Shunt Fraction

$sO_{2\ PC}$ (%)	sO_2 art (%)	sO_2 ven (%)	Shunt fraction (%)
100	99	60	2.5 (normal)
100	94	55	13 (elevated)

$sO_{2\ PC}$: Oxygen saturation of pulmonary capillary blood.

heart rate, and electrocardiogram, a more accurate method is by a pulmonary artery catheter with a thermodilution tip that measures blood flow (cardiac output). A known quantity of cold solution is injected, with the change in blood temperature measured and related to blood flow.

A simple laboratory test that provides a sensitive index of the overall state of oxygen delivery and oxygen utilization is measurement of blood lactate *(35)*.

COLLECTION AND HANDLING OF SAMPLES FOR BLOOD GAS ANALYSIS

Use of Arterial vs. Venous Blood for Blood Gas and Acid-Base Measurements

Perhaps more than any other analytes, pO_2 and pCO_2 change markedly from arterial to venous blood, whereas pH changes slightly. Arterial blood is nearly always preferred over venous blood for blood gas analysis of pH, pCO_2, and pO_2 for these reasons:

- Arterial pO_2 indicates the ability of the lungs to oxygenate, or equilibrate, the blood with alveolar air.
- Arterial blood provides an index of the oxygen and nutrients that will be provided to the tissues and cells.

Because typical venous blood collected from an arm vein represents oxygen metabolism only in the arm, venous blood representing oxygen metabolism of the entire body is sometimes needed. This requires that "mixed" venous blood be collected, which exists in the pulmonary artery and can be collected with a Swan-Ganz catheter *(14)*. Both mixed venous and arterial blood are needed when determining parameters such as P50 and VO_2.

Collection and Handling of Blood

Blood collected for blood gas analysis is highly susceptible to changes in pO_2. Anaerobic conditions during collection and handling are essential because room air has a pCO_2 of nearly 0 and a pO_2 of ~150 mmHg. These are the factors that must be controlled:

- Removal of all air bubbles
- Use of the proper anticoagulant
- Appropriate use of plastic syringes (glass syringes rarely used)
- The temperature of storage before analysis
- The length of delay between collection and analysis of blood

The complete removal of all air bubbles is especially important before sending blood in a syringe by pneumatic transport, which can markedly affect pO_2 *(36)*; on the other hand, pCO_2 is almost unaffected by air bubbles.

Liquid heparin at <10% (vol./vol.) of the volume of blood should have little effect on pH, pCO_2, or pO_2 *(16)*. However, the use of liquid heparin will dilute other constituents in blood, such as electrolytes and glucose, which are often analyzed simultaneously in many current blood gas and electrolyte analyzers. Therefore, only dry heparin should be used as an anticoagulant.

Although plastic syringes are used for nearly all blood gas measurements, they have a potential disadvantage due to their ability to absorb oxygen *(37)*. When stored in ice water, because of the increased oxygen affinity of Hb at cold temperatures, blood can absorb oxygen dissolved within the wall of the syringe and transmitted through the plastic. This effect is most pronounced in samples with a pO_2 of ~100 mmHg and above; that is, when Hb is already nearly fully saturated with oxygen and is unable to buffer any added O_2. A pO_2 of 100 mmHg may increase by 8 mmHg during 30 minutes of storage on ice.

When Hb is less saturated (for example, at a pO_2 of 60 mmHg), it is better able to buffer the additional oxygen, with little or no measurable change in pO_2. Because glass is not permeable to O_2, pO_2 is not affected in blood stored in glass syringes in ice.

In general, storage of blood in plastic syringes at room temperature is acceptable if analysis occurs within 15 minutes. When blood is stored at ambient temperature for >30 minutes, cellular metabolism can decrease pO_2, and ice storage may be necessary. Storage for 15 minutes at ambient temperature (22–24 °C; 72–75 °F) changes pO_2 by <1 mmHg even at a pO_2 of 100 mmHg *(24)*. In most samples at ambient temperature (22–24 °C; 72–75 °F), pCO_2 changes <1 mmHg, and pH changes <0.01 unit. However, samples from patients with extreme leukocytosis can dramatically change pH, pCO_2, and pO_2 (and glucose and lactate) when stored at room temperature. These samples must be analyzed as soon as possible.

CORD BLOOD GASES

Blood gas values on blood collected from the umbilical cord at birth can provide valuable data for assessing the respiratory and metabolic status of the newborn. They are more objective than apgar scores at identifying hypoxia and acidosis in the neonate that may occur from asphyxia, respiratory distress, or other conditions. In most cases, blood from the umbilical artery should be collected because it contains blood returning from the fetus to the placenta and more accurately represents the condition of the neonate. The placenta performs both "pulmonary" gas exchange and "renal" functions such that blood from the umbilical vein is similar to that of maternal circulation. Some studies have concluded that collection of both umbilical arterial and venous blood provides clinically useful information on the pathogenesis of acidosis *(38)*.

When the fetus cannot remove sufficient CO_2 via the placenta, the pCO_2 increases leading to a "respiratory" acidosis. If O_2 exchange is inadequate (from multiple causes), a metabolic acidosis results. Here are some critical results for cord blood:

Foundations of Blood Gases

1. pH less than 7.0 in arterial cord blood
2. pCO_2 difference between arterial cord blood and venous cord blood >25 mmHg
3. pH difference between arterial and venous cord blood >0.12

The relevance of pO_2 results on cord blood is not clearly established. Reference ranges for both arterial and venous cord blood have been published (38) and are summarized in Table 1-16.

TABLE 1-16. Reference Ranges for Arterial and Venous Cord Blood

Analyte	Arterial cord blood reference range	Venous cord blood reference range
pH	7.14–7.42	7.22–7.44
pCO_2 (mmHg)	34–78	30–63
pO_2 (mmHg)	3–40	12–43

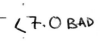

REFERENCES

1. Grogono AW. Acid-base tutorial: acid-base history. http:/www.acid-base.com (Accessed Dec 2008).
2. Story DA. Bench-to-bedside review: a brief history of clinical acid-base. Crit Care 2004;8:253–258.
3. Astrup A, Severinghaus JW. The history of blood gases, acids, and bases. Copenhagen: Munksgaard, 1986.
4. Grogono AW. Acid-base tutorial terminology: base excess. http:/www.acid-base.com (Accessed Dec 2008).
5. Brandis K. Acid-base pHysiology: the anion gap. http://www.anaesthesiaMCQ.com/AcidBaseBook (Accessed Jan 2009).
6. Iberti TJ, Leibowitz AB, Papadakos PJ, Fischer EP. Low sensitivity of the anion gap as a screen to detect hyperlactatemia in critically ill patients. Crit Care Med 1990;18:275–7.
7. Story DA, Morimatsu H, Bellomo R. Strong ions, weak acids, and base excess: a simplified Fencl-Stewart approach to clinical acid-base disorders. Brit J Anaesthesia 2004;92:54–60.
8. Grogono AW. Acid-base tutorial: Stewart's strong ion difference. http:/www.acid-base.com (Accessed Dec 2008).
9. Scott MG, LeGrys VA, Klutts JS. Electrolytes and blood gases. In: Burtis CA, Ashwood ER, Bruns DE, ed. Tietz textbook of clinical chemistry and molecular diagnostics, 4th ed. St. Louis: Elsevier Saunders, 2006:983–1018.
10. Toffaletti J, Zijlstra WG. Misconceptions in reporting oxygen saturation. Anesth Analg 2007;105:S5–9.
11. Robergs RA, Ghiasvand F, Parker D. Biochemistry of exercise-induced metabolic acidosis. Am J Physiol Regul Integr Comp Physiol 2004;287:R502–16.
12. Ehrmeyer SS, Shrout JB. Blood gases, pH, and buffer systems. In: Bishop ML, Duben-Engelkirk JL, Fody EP, eds. Clinical chemistry: principles, procedures, correlations, 3rd ed. Philadelphia: JB Lippincott, 1996:237–53.

13. Kraut JA, Madias NE. Serum anion gap: its uses and limitations in clinical medicine. Clin J Am Soc Nephrol 2007;2:162–74.
14. Shapiro BA, Peruzzi WT, Templin R. Clinical applications of blood gases, 5th ed. St. Louis: CV Mosby, 1994.
15. Brandis K. Acid-base pHysiology: metabolic alkalosis—compensation. http://www.anaesthesiaMCQ.com/AcidBaseBook (Accessed Jan 2009).
16. Cornelissen PJH, van Woensel CLM, van Oel WC, de Jong PA. Correction factors for hemoglobin derivatives in fetal blood, as measured with the IL282 CO-Oximeter. Clin Chem 1983;29:1555–6.
17. Robinson JM, Lancaster JR Jr. Hemoglobin-mediated, hypoxia-induced vasodilation via nitric oxide. Am J Resp Cell Molecular Biol 2005;32:257–61.
18. Grogono AW. Acid-base tutorial: metabolic acidosis and alkalosis. http:/www.acid-base.com (Accessed Dec 2008).
19. Brandis K. Acid-base pHysiology: metabolic acidosis—causes. http://www.anaesthesiaMCQ.com/AcidBaseBook (Accessed Jan 2009).
20. University of Connecticut. Acid base online tutorial. Compensatory responses: metabolic acidosis. http://fitsweb.uchc.edu/student/selectives/TimurGraham/Compensatory_responses_metabolic_acidosis.html (Accessed Dec 2008).
21. Brandis K. Acid-base pHysiology: metabolic acidosis—correction. http://www.anaesthesiaMCQ.com/AcidBaseBook (Accessed Jan 2009).
22. Brandis K. Acid-base pHysiology: metabolic alkalosis—causes. From http://www.anaesthesiaMCQ.com/AcidBaseBook (Accessed Jan 2009).
23. University of Connecticut. Acid base online tutorial. Compensatory responses: metabolic alkalosis. http://fitsweb.uchc.edu/student/selectives/TimurGraham/Compensatory_responses_metabolic_alkalosis.html (Accessed Dec 2008).
24. Adrogue HJ, Madias NE. Management of life-threatening acid-base disorders. New Engl J Med 2005;338:107–11.
25. Brandis K. Acid-base pHysiology: respiratory acidosis—causes. http://www.anaesthesiaMCQ.com/AcidBaseBook (Accessed Jan 2009).
26. Narins RG, ed. Maxwell and Kleeman's clinical disorders of fluid and electrolyte metabolism, 5th ed. New York: McGraw-Hill, 1994.
27. Respiratory acidosis—Wikipedia. http://en.wikipedia.org/wiki/Respiratory_acidosis. (Accessed Dec 2008).
28. Brandis K. Acid-base pHysiology: respiratory acidosis—correction. http://www.anaesthesiaMCQ.com/AcidBaseBook (Accessed Jan 2009).
29. Diagnosing mixed acid-base disorders. http://www.lakesidepress.com/pulmonary/ABG/MixedAB.htm (Accessed Dec 2008).
30. Brandis K. Acid-base pHysiology: respiratory alkalosis—correction. http://www.anaesthesiaMCQ.com/AcidBaseBook (Accessed Jan 2009).
31. Walmsley RN, White GH. Mixed acid-base disorders. Clin Chem 1985;31:321–5.
32. Brandis K. Acid-base pHysiology: the delta ratio. http://www.anaesthesiaMCQ.com/AcidBaseBook (Accessed Jan 2009).
33. Walmsley RN, White GH. A guide to diagnostic clinical chemistry, 3rd ed. Oxford: Blackwell Scientific, 1994:79–127.
34. Williams AJ. ABC of oxygen: assessing and interpreting arterial blood gases and acid-base balance. British Med J 1998;317:1213–6.
35. Tuchschmidt J, Oblitas D, Fried JC. Oxygen consumption in sepsis and septic shock. Crit Care Med 1991;19:664–71.

36. Astles JR, Lubarsky D, Loun B, Sedor FA, Toffaletti JG. Pneumatic transport exacerbates interference of room air contamination of blood gas samples. Arch Pathol Lab Med 1996;120:642–7.
37. Mahoney TJ, Harvey JA, Wong RJ, Van Kessel AL. Changes in oxygen measurements when whole blood is stored in iced plastic or glass syringes. Clin Chem 1991;37:1244–8.
38. Fouse BL. Reference range evaluation for cord blood gas parameters. June 2002. www.acutecaretesting.org (Accessed April 2009).

Chapter 2

Calcium, Magnesium, and Phosphate

Three electrolytes—calcium (Ca), magnesium (Mg), and phosphate (PO_4)—have a prominent role in diagnostic tests. This chapter reviews the historical and diagnostic significance of these electrolytes in human health, and provides information on their physiology, regulation, distribution, and related conditions. The text also examines evaluation and interpretation of measurements, and provides reference ranges, for these analytes.

HISTORY AND SIGNIFICANCE

In 1883, Sydney Ringer showed that **calcium** was essential for myocardial contraction, and in 1934, Franklin McLean and Baird Hastings showed that the ionized calcium concentration was proportional to the amplitude of frog heart contraction, whereas protein-bound and citrate-bound calcium had no effect on contractions. After many attempts were made to develop improved methodology for measuring ionized calcium, reliable ion-selective electrodes (ISE) were finally developed and are now available.

In 1695, from well water in Epsom, England, Dr. Nehemiah Grew prepared Epsom salts, a name still given to **magnesium** sulfate. The biological significance of magnesium as a constituent of plants has been known since the 18th century, with magnesium ion an essential component of chlorophyll. Measurements of ionized magnesium have become available, with reliability not yet to the standard of ionized calcium.

Symptoms of **phosphate** depletion have been described through the years, from the times of the ancient Romans to observations by veterinarians on livestock.

The electrolytes calcium, magnesium, and phosphate have many roles in the structure and function of bone, the function of membranes, and the activation of hundreds of enzymes involved in genetic regulation, muscle contraction, and energy utilization. Parathyroid hormone and vitamin D regulate the distribution of these ions among bone, soft tissues, and extracellular fluids (ECFs). As with most electrolytes, the kidneys have a major role in regulating the concentration of these ions in blood.

Other than for diagnosing parathyroid dysfunction and hypercalcemia of malignancies, the usefulness of these electrolytes in diagnostic tests was of only moderate interest through the 1970s. Over the past 20–25 years, however, a prominent role has evolved for these tests in monitoring patients, including neonates, during surgery and critical care. The role of intracellular calcium ions and the protection by magnesium ion in reperfusion injury after ischemia, the potentially fatal complications from hypomagnesemia in cardiovascular disease, and the importance of adequate phosphate for regulation of oxygen delivery and energy utilization are some examples of the vital importance of these electrolytes in critically ill patients.

Understanding the regulation of calcium and other electrolytes by parathyroid hormone (PTH), vitamin D, calcitonin, and possibly parathyroid hormone-related protein (PTHRP) has helped in the differential diagnosis of many calcium disorders. The regulation of magnesium is less well understood, with PTH having a major role and hormones such as vitamin D, aldosterone, and insulin also involved.

CALCIUM

Physiology

The balance between the very low concentration of calcium in cytoplasm and the relatively high concentrations in organelles and ECFs controls many neuromuscular and secretory processes. Because decreased ionized calcium can impair myocardial function, maintaining ionized calcium within a clinically acceptable concentration range is especially important in patients under critical care *(1)*.

The inward flow of calcium ions into cardiac cells helps control cardiac rhythm by binding to contractile proteins in myocardial cells, which initiates the contractile process. The rate of flow of calcium ions into smooth muscle cells influences the tension of arterioles that regulate blood pressure. Calcium channels in the cell membrane regulate this flow by opening when the membrane is depolarized. Cardiotropic drugs such as epinephrine and isoproterenol facilitate transport of calcium ions through these channels promoting contraction, whereas acetylcholine hinders the transport of calcium ions *(2)*.

Calcium ions enter cells through calcium channels. This triggers the release of calcium ions from both the sarcoplasmic reticulum and the inner cell membrane. ATPases remove the excess calcium ions from inside the cell and transport calcium ions back into the sarcoplasmic reticulum for future release. Two types of calcium channels on the cell membrane are known to regulate the flow of calcium: a voltage-dependent and a phosphorylation-dependent channel. The voltage-dependent channel opens when the membrane is depolarized, whereas the phosphorylation-dependent channel requires a protein kinase activated by cyclic adenosine monophosphate (cAMP) to open. Several substances are known to affect these channels: the cardiotropic drugs epinephrine and isoproterenol facilitate transport of calcium ions;

acetylcholine hinders transport of calcium ions (3); magnesium ion stabilizes these channels.

Calcium ions are also important as second messengers in controlling the secretion of many hormones, such as insulin, aldosterone, vasopressin, and renin. After stimulation at the cell surface by a specific molecule (first messenger), the inward flux of calcium ions (second messenger) initiates production of a hormone.

Symptoms of hypocalcemia are often manifested as cardiovascular disorders such as cardiac insufficiency, hypotension, and arrhythmias. Hypocalcemia can also cause irregular muscle spasms, called tetany. The rate of fall in ionized calcium initiates symptoms as much as the absolute concentration of ionized calcium. Hypercalcemia can cause a general lethargy in patients and can affect multiple organ systems. These can include changes in electric conduction pathways of the heart, GI smooth muscle relaxation causing constipation and nausea, and kidney effects that can lead to dehydration and nephrolithiasis (1).

Regulation in the Blood

Three hormones (PTH, vitamin D, and calcitonin, although the role of calcitonin is less clear) are regarded as the principal regulators of serum calcium because they alter their secretion rate in response to changes in ionized calcium. The actions of these hormones are shown in Figure 2-1. A molecule with PTH-like actions, called PTH-related peptide (PTHrP), appears to have

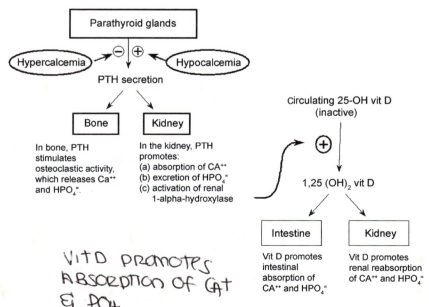

FIGURE 2-1. Hormonal responses to hypercalcemia and hypocalcemia. Vit D, vitamin D; $HPO_4^=$, HPO_4^{2-}.

hypercalcemic actions similar to those of PTH, and may be especially important in hypercalcemia of malignancy.

When calcium levels in blood fall, cells in the parathyroid glands respond by synthesizing and secreting PTH, which acts synergistically with vitamin D to increase blood calcium levels. The parathyroid glands respond rapidly to a decrease in ionized calcium concentration in blood, with a fourfold increase in PTH secretion stimulated by a 5% decrease in ionized calcium *(3,4)*, which stops when the ionized calcium increases *(5)*. PTH is synthesized in the parathyroid gland as a precursor molecule of 115 amino acids. After cleavage in the gland, intact PTH (1–84) is secreted into the blood, where it circulates as intact hormone and amino-terminal, carboxy-terminal, and midmolecule fragments of varying biological activity. Both intact PTH and amino-terminal PTH have biological activity by interacting with PTH receptors on bone and renal cell membranes *(6)*. In bone, PTH activates osteoclasts to release cytokines that enhance breakdown of bone to release calcium and phosphate into the ECF. In kidneys, PTH conserves calcium by increasing tubular reabsorption of calcium ions and lowers phosphate by inhibiting tubular reabsorption of phosphate. PTH also activates a specific hydroxylase enzyme in the kidney that increases production of active vitamin D.

Inactive forms of vitamin D_3 are obtained either from the diet or from exposure of skin to sunlight. Light converts precursors in the diet to either vitamin D_2 or vitamin D_3. These are converted in the liver to 25-hydroxycholecalciferol (25-OH-D_3), then activated by an enzyme (1α-hydroxylase) in the kidney to form 1,25-dihydroxycholecalciferol [1,25-$(OH)_2$-D_3], the biologically active form. PTH promotes the conversion of inactive to active vitamin D, by activating the production of the 1α-hydroxylase enzyme, and phosphate inhibits production. Active vitamin D

- Enhances activity of calcium pumps and calcium channels in intestinal cells to increase calcium and phosphate absorption in the intestine.
- Has a short-term enhancement of PTH-activated bone resorption to release calcium into blood.
- Acts with PTH to minimize renal excretion of calcium.
- Contributes to the negative feedback regulation of PTH through specific receptors for 1,25-$(OH)_2$-D_3 on the parathyroid glands.

For a variety of reasons, the frequency of vitamin D deficiency is rising, which has caused a dramatic increase in vitamin D testing.

If calcium levels in blood increase, the medullary cells in the thyroid gland release calcitonin, which exerts its calcium-lowering effect by inhibiting the osteoclasts in bone. This in turn promotes deposition of calcium into bone, which lowers calcium levels in blood. Calcitonin may cause this effect by inhibiting the actions of both PTH and vitamin D on these cells in the thyroid. Calcitonin may not contribute to normal calcium homeostasis, because (a) persons without thyroid glands are still able to regulate their calcium levels in blood, and (b) calcitonin does not change in response to small (1–2%) increases in ionized calcium. A study on hemodialysis patients showed that ionized calcium had to increase by ~10% to stimulate a calcitonin response (7).

PTHrP is a peptide with some functions similar to PTH (8). PTHrP has been detected in many patients with humoral hypercalcemia of malignancy. NH_2-terminal PTHrP (1–36) is equipotent to PTH both in promoting renal calcium reabsorption and in stimulating bone resorption. In the normal fetus, PTHrP (1–36) apparently activates production of 1,25-$(OH)_2$ vitamin D in the kidney, and PTHrP (1–84) promotes placental transport of calcium into the fetus. It has a major pathological role in mediating bone destruction and hypercalcemia in certain tumors. Inhibiting its action in this condition may become an important therapy (8). However, it has no major physiologic role in normal calcium homeostasis, possibly because current assays are barely able to detect PTHrP at normal plasma concentrations (~1 pmol/L).

Distribution in Cells and Blood

More than 99% of calcium in the body is in bone as hydroxyapatite, a complex molecule of calcium and phosphate. The remaining 1% is mostly in the blood and other ECFs. The amount of calcium is lower in cells than in blood, and the concentration of free ionized calcium in the cytosol of cardiac or smooth muscle cells is much lower (1/5000 to 1/10,000 as much). Maintenance of this large gradient ensures the rapid inward flux of calcium ions necessary to trigger muscle contraction.

Calcium in blood is distributed among several forms. About 45% circulates as free calcium ions, 40% is bound to protein (mostly albumin), and 15% is bound to anions such as HCO_3^-, citrate, phosphate, and lactate (9). As shown in Figure 2-2, bound forms of calcium are in equilibrium with the free calcium ions. This distribution can change somewhat in many diseases, but major changes may occur during surgery or in critically ill patients because of changes in citrate, HCO_3^-, lactate, and phosphate. These are the principal

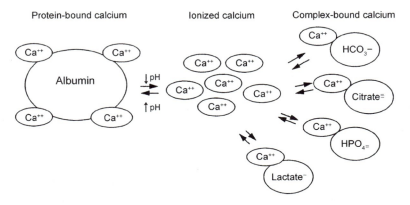

FIGURE 2-2. Equilibria between protein-bound, complex-bound, and ionized calcium. Ca^{++}, Ca^{2+}; $HPO_4^=$, HPO_4^{2-}; $citrate^=$, $citrate^{3-}$.

reasons why ionized calcium cannot be reliably calculated from total calcium measurements in acutely ill individuals. As shown in Figure 2-2, because pH affects calcium binding to albumin, it also affects the concentration of ionized calcium. As a rough guide, each 0.1 pH unit decrease will increase ionized calcium by ~0.05 mmol/L (0.2 mg/dL) *(10)*.

Evaluation of Calcium Measurements

Most calcium disorders can be related to

- A defect in parathyroid function or vitamin D metabolism
- Malignancy
- Renal disease
- Excess administration of citrate, calcium, or saline

Calcium measurements appear to be most useful in the following five clinical areas:

1. The recognition of hypercalcemia due to malignancy
2. The recognition of hypercalcemia due to parathyroid disease
3. In renal diseases
4. In monitoring of neonates, during and after surgery
5. In critically ill patients

Causes of Hypocalcemia

Decreased ionized calcium concentrations in blood can cause cardiac insufficiency, congestive heart failure, arrhythmias, and neuromuscular irritability, which may become clinically apparent as irregular muscle spasms and tetany, confusion, and seizures. Studies have shown that the rate of fall of ionized calcium in blood initiates tetany as much as the absolute decrease in concentration of ionized calcium *(11)*.

Hypocalcemia occurs most commonly in the following patients:

- Those with parathyroid gland insufficiency from parathyroid or thyroid surgery
- defect of the calcium-sensing receptor, either genetic or autoimmune
- end-organ resistance to PTH (pseudohypoparathyroidism)
- Those receiving citrated blood products during major surgery
- Patients with 1,25-(OH)$_2$ vitamin D deficiency or inhibition of 1α-hydroxylase due to hyperphosphatemia in renal disease
- Neonates
- Patients with magnesium deficiency
- Patients with pancreatitis

These and other frequent causes of hypocalcemia are presented in Table 2-1. Hypocalcemia due to either surgical removal of excess parathyroid tissue in parathyroid surgery or parathyroid hypofunction after thyroid

Calcium, Magnesium, and Phosphate

[handwritten: Ca ~ 9-10]

[handwritten: ALBUMIN ~ 3-5]

TABLE 2-1. Most Frequent Causes of Hypocalcemia

- Chronic renal failure
- Hyperphosphatemia
- Chelators: citrate, heparin, albumin, EDTA
- Hypomagnesemia
- Vitamin D deficiency or defect
- Hypoparathyroidism: postsurgical or pseudohypoparathyroidism
- Illnesses in neonates
- Pancreatitis
- Sepsis
- Burns
- Hypermagnesemia

[handwritten notes: WHEN PT IS HYPOALBUMIN ↓ CORRECTED Ca = TOT SERUM Ca + (0.8 x [4 - ALBUMIN]) — Ca+ fall 0.8 mg/dL for a 1 g/dL fall in ALBUMIN. ABN ALBUMIN THINK LIVER OR KIDNEY ISSUE]

[handwritten: ↓ TOTAL SERUM Ca+ BUT NORMAL IONIZED Ca+ → PSEUDOHYPOCALCEMIA.]

surgery is usually transient (lasting <5 days) unless surgery has removed too much parathyroid tissue or has interfered with parathyroid blood supply. Intraoperative measurements of PTH have become a standard of practice during parathyroid surgery to assure that the appropriate amount of parathyroid tissue is removed *(12)*. It is also useful to monitor serum calcium after neck surgery, until a rise in the ionized calcium indicates recovery of the parathyroid gland *(2,13)*.

The sequence of laboratory tests to be ordered after confirmation of hypocalcemia as a low ionized calcium is shown in Figure 2-3. Magnesium is an appropriate test to order, because magnesium deficiency inhibits the secretory transport of PTH across the membrane.

Although total calcium is often the initial laboratory test used to detect hypocalcemia, total calcium may not agree with ionized calcium. As emphasized many years ago *(14)*, ionized calcium cannot be accurately calculated from total calcium. For example, it is not uncommon for a patient with a low total calcium and low albumin (who might be presumed to be normocalcemic) to have a low ionized calcium. More recent reports have re-confirmed that ionized calcium cannot be accurately predicted by correcting the total calcium concentration based on albumin concentration *(15–17)*.

Most current instruments provide an ionized calcium parameter as ionized calcium adjusted to a pH of 7.4. This term was introduced as a convenience for samples that have been exposed to air. Because pH-corrected results may be misleading in some patients, actual ionized calcium should be measured on anaerobically collected blood.

Renal diseases

Patients with renal disease often have altered concentrations of calcium, phosphate, albumin, magnesium, and H^+ (pH). Both hyperphosphatemia and hypocalcemia are often features of renal disease, most likely due to complex alterations in the production and/or response to both PTH and vitamin D.

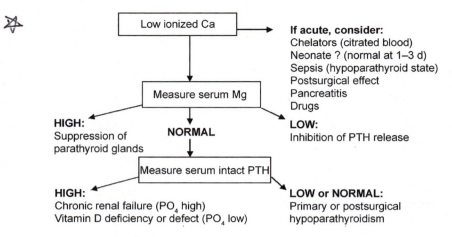

FIGURE 2-3. Use of laboratory tests in evaluation of hypocalcemia.

In chronic renal disease, secondary hyperparathyroidism frequently develops to compensate for hypocalcemia caused by diminished vitamin D production due to inhibition of renal 1α-hydroxylase. Monitoring and controlling ionized calcium in renal disease and renal dialysis may avoid problems due to hypocalcemia, such as osteodystrophy, unstable cardiac output, or unstable blood pressure, or problems arising from hypercalcemia such as soft tissue calcifications *(18)*.

Neonatal monitoring

Ionized calcium concentrations in the blood of apparently normal neonates are high at birth, rapidly decline by 10–20% after 1–3 days, then stabilize at concentrations slightly higher than in adults after ~1 week *(19)*. Such events in healthy infants appear to be a normal physiological stimulus to activate parathyroid gland function. Another form of hypocalcemia that develops after ~1 week is associated both with hyperphosphatemia from a high phosphate intake from milk and with hypomagnesemia caused by decreased intestinal absorption of magnesium *(2)*.

Other factors related to the incidence and severity of hypocalcemia in neonates include *(2)*:

- Prematurity
- Maternal diabetes mellitus
- Complications during delivery
- Birth asphyxia

If hypocalcemia is severe or prolonged or accompanied by seizures, hypotension, hypoglycemia, or sepsis, it may be life threatening. In such patients, moderately large changes in the concentration of ionized calcium may occur as calcium is lost rapidly by the infant and not readily reabsorbed. Several possible etiologies have been suggested *(19)*:

- Abnormal PTH and vitamin D metabolism
- Hypercalcitoninemia
- Hyperphosphatemia
- Hypomagnesemia

Surgery

During surgery or intensive care, adequate calcium concentrations enhance cardiac output and maintain blood pressure. Monitoring and adjusting calcium concentrations is important in open heart surgery when the heart is restarted, because normalizing ionized calcium by administering calcium is a prudent action to prevent the cardiac alterations associated with hypocalcemia *(1)*. Monitoring ionized calcium is important during liver transplantation because large volumes of citrated blood are given at a time when liver function (the major organ for metabolizing citrate) is compromised or absent.

One report suggests that if calcium is administered and carefully titrated to maintain a normal ionized calcium concentration in blood, adverse side effects can be avoided *(20)*. The same report also suggested that hypocalcemia may have a protective mechanism against cell death during cellular hypoxia.

Hypoparathyroidism

Hypoparathyroidism may be acquired as a genetic or autoimmune disorder or, more commonly, may arise after thyroid or parathyroid surgery. Postsurgical hypoparathyroidism is usually transient unless surgery has removed too much parathyroid tissue or interfered with parathyroid gland blood supply. Intraoperative PTH measurements are commonly done to ensure the appropriate amount of parathyroid tissue has been removed *(12)*.

In pseudohypoparathyroidism, renal cells apparently do not respond to PTH, which explains the elevated circulating concentrations of PTH. The combined effects of PTH resistance and hyperphosphatemia can lead to suppression of 1α-hydroxylase activity, producing a deficiency of active vitamin D and hypocalcemia *(21)*.

Hungry bone syndrome

Hungry bone syndrome refers to hypocalcemia caused by the deposition of calcium into bone at an accelerated rate during healing of osteodystrophies. This syndrome is sometimes seen after surgical correction of long-standing secondary hyperparathyroidism by partial parathyroidectomy. It may also occur from osteoblastic metastases in various malignancies.

Critically ill patients

In critically ill patients, that is, those with sepsis, thermal burns, renal failure, and/or cardiopulmonary insufficiency, hypocalcemia may occur in as many as 70% of cases. These patients frequently have abnormal acid-base regulation, diminished protein and albumin, and ionized hypocalcemia, which may directly affect mean arterial pressure in intensive care unit patients *(22)*. The inflammatory response likely plays a role in the development of these conditions.

Hypomagnesemia

As hypomagnesemia in hospitalized patients has become more frequent, chronic hypomagnesemia has also become recognized as a frequent cause of hypocalcemia. Hypomagnesemia may cause hypocalcemia by

- Inhibiting the glandular secretion of PTH across the parathyroid gland membrane *(23)*.
- Impairing PTH action at its receptor site on bone *(2)*.
- Causing vitamin D resistance *(24)*.

Pancreatitis

About 40–75% of patients with acute pancreatitis develop hypocalcemia. While the standard mechanism is often cited as precipitation of calcium ions by free fatty acids generated during acute pancreatitis, the mechanism is certainly more complex. Parathyroid dysfunction, calcitonin release, and endotoxin release may play roles as well. A consistent cause appears to be diminished secretion of PTH, which may either be low or inappropriately normal for the degree of hypocalcemia *(21)*.

Causes of Hypercalcemia

Because the clinical symptoms of hypercalcemia are relatively nonspecific (confusion, muscle weakness, vomiting, arrhythmia), the diagnosis must be confirmed and evaluated by laboratory measurements. A total calcium measurement >3 mmol/L (>12 mg/dL) does not usually require confirmation with an ionized calcium measurement. However, a slightly to moderately elevated total calcium should be confirmed with ionized calcium, if available. Interpretation of total calcium results "corrected for albumin" is a clinical practice based on skilled guesswork, and it may be quite misleading when monitoring plasma calcium in critically ill patients *(15–17)*. The sequence of laboratory tests to be ordered after detection of hypercalcemia is shown in Figure 2-4.

Malignancy and primary hyperparathyroidism (HPTH) are the most common causes of hypercalcemia, accounting for 80–90% of all hypercalcemic patients *(25)*. Hypercalcemia due to malignancy is more likely in a hospital population, whereas primary HPTH is more commonly found in the outpatient population.

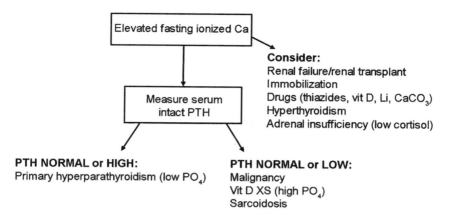

FIGURE 2-4. Use of laboratory tests in evaluation of hypercalcemia. vit D, vitamin D; XS, excess.

Much less common causes of hypercalcemia include the following, which may often be ruled out by clinical history:

- Prolonged immobilization for >1 week [this can elevate ionized calcium by 0.2–0.5 mmol/L (0.8–2.0 mg/dL)] *(26)*
- Drug therapy from thiazide diuretics, lithium, vitamin D, or antacids
- Milk alkali syndrome
- Hyperthyroidism
- Adrenal insufficiency
- Familial hypocalciuric hypercalcemia (FHH)
- Sarcoidosis
- Acute renal failure

FHH is a genetic disorder characterized by low urinary excretion of calcium in a setting of hypercalcemia. Adrenal insufficiency apparently causes hypercalcemia by creating a deficiency of cortisol. Because cortisol inhibits osteoclast activity and antagonizes the action of vitamin D, a deficiency of cortisol promotes osteoclastic activity and vitamin D action, both of which lead to hypercalcemia.

If other causes of hypercalcemia are ruled out, as shown in Figure 2-4, the clinician must differentiate between primary HPTH and hypercalcemia due to occult malignancy. A history of chronic stable mild hypercalcemia without symptoms or weight loss is consistent with primary hyperparathyroidism. On the other hand, an acutely rising calcium, weight loss, and fever suggest malignancy as the cause of hyperparathyroidism. Carcinomas of the bronchus, breast, head and neck, urogenital tract, and multiple myeloma account for 75% of the hypercalcemia in malignancy *(25)*.

Typical results of laboratory tests in differentiating primary hyperparathyroidism from malignancy are shown in Table 2-2 *(27)*. In general, ionized calcium measurements are elevated in 90–95% of cases of HPTH, while total calcium is elevated in 80–85% of cases *(28)*.

TABLE 2-2. Interpretation of Laboratory Tests in Differentiating Primary Hyperparathyroidism from Malignancy

Test	Favors HPTH	Favors malignancy
Total Ca	<3.13 mmol/L (<12.5 mg/dL)	>3.13 mmol/L (>12.5 mg/dL)
Serum Cl[a]	>103 mmol/L (>103 mEq/L)	<103 mmol/L (<103 mEq/L)
Intact PTH	Elevated (or high normal)	Suppressed
Serum PO_4[a]	Normal to low	Variable
Hematocrit	Normal	Low
Urine Ca	High	Very high
1,25-$(OH)_2$ vitamin D	High	Low
PTHrP	Normal or undetected	Elevated

[a]Some clinicians also use a Cl:PO_4 ratio as evidence for HPTH. Usually, Cl:PO_4 is >102 with SI units (Cl^- and PO_4 in mmol/L) and >33 with conventional units (Cl^- in mEq/L, PO_4 in mg/dL) in HPTH. Although often high, serum phosphate may also be low in cases with PTHrP production.

In the patient with HPTH, removal of excess hyperplastic parathyroid tissue may be necessary to correct hypercalcemia. During surgical removal of excess parathyroid tissue, rapid measurement of PTH in blood during the operation has proven very helpful in determining whether sufficient parathyroid tissue has been removed *(12,29)*.

In some patients with secondary HPTH, successful renal transplantation removes the hypocalcemic stimulation to the parathyroid gland. In some cases of prolonged secondary HPTH, the long-hypertrophied parathyroid glands develop autonomous PTH secretion, which leads to tertiary HPTH. This condition resembles primary HPTH and may require surgical removal of the appropriate amount of parathyroid tissue, which may be more than in primary HPTH *(30)*.

Interpretation of Calcium and PTH Measurements

In differentiating various causes of hypocalcemia and hypercalcemia, concurrent measurements of PTH and calcium can be very helpful *(31)*. Figure 2-5 shows the relationship between total calcium and intact PTH expected for primary hypo- and hyperparathyroidism, secondary hyperparathyriodism, and hypercalcemia of malignancy.

Although several assays have been developed for measuring various forms of the PTH molecule, the intact PTH assay appears to have the best combination of clinical specificity and sensitivity. It detects the entire biologically active PTH molecule of 84 amino acids, which is the major form of PTH secreted from the gland. Intact PTH has a relatively short half-life in plasma (5 minutes). The NH_2-terminal PTH assays detects the biologically active amino terminal (1–34 amino acid) fragment of PTH and intact PTH, both of which have relatively short half-lives in plasma.

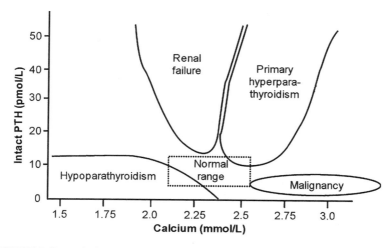

FIGURE 2-5. Relationship of total calcium and PTH in calcium disorders.

The midmolecule and COOH-terminal PTH assays detect PTH fragments (including intact PTH) that contain the midregion or COOH-terminal end of the PTH molecule. Because many PTH fragments have a longer half-life in plasma (30 minutes) than the intact hormone, these assays are the most sensitive for detecting hyperactivity of the parathyroid gland. However, because they cross-react with other peptides in serum, causing PTH results to be inappropriately high (or normal) in patients with hypoparathyroidism or hypercalcemia of malignancy, these assays are seldom used for routine PTH testing. Decreased glomerular filtration also tends to spuriously elevate these PTH results by prolonging the life of these fragments in the circulation.

Proper Collection and Handling of Samples

When collecting samples for either total or ionized calcium measurements, tourniquet application should be brief (<1 minute) before a specimen is collected. The combined effects of hemoconcentration (causing hyperproteinemia) and localized lactate and acid production can alter total calcium by as much as 10% *(28)* and ionized calcium by ~2–3% *(32)*. Because loss of CO_2 will increase pH, all samples for ionized calcium measurements must be collected anaerobically. Metabolic activity of cells during storage affects both pH and ionized calcium. Blood should be centrifuged within 1–2 hours to prevent acidosis from affecting the ionized calcium concentration. Because of dilutional effects, no liquid heparin anticoagulants should be used. Most heparin anticoagulants (sodium, lithium) partially bind to calcium and lower ionized calcium concentrations. However, syringes containing dry heparin products are available that essentially eliminate the interference by heparin *(33,34)*. These include heparin that has been "balanced" with electrolytes that minimize the binding of calcium in the blood sample to the heparin, and minimal heparin that is dispersed in a soluble inert carbohydrate "web." Furthermore, incomplete filling of syringes containing these heparins has minor effects on the ionized calcium and magnesium concentration *(34)*.

For analysis of total calcium in urine, an accurately timed urine collection is preferred. The urine should be acidified with 6 mol/L HCl, with ~1 mL of the acid added for each 100 mL of urine.

Reference Ranges for Calcium

For total calcium, the reference range varies slightly with age. In general, calcium concentrations are higher through adolescence while bone growth is most active. Ionized calcium concentrations can change rapidly during the first 1–3 days of life. After this, they stabilize at relatively high concentrations, with a gradual decline through adolescence (see Table 2-3).

Urine reference ranges for total calcium vary with diet. Patients on an average diet should excrete 50–300 mg/d. There is no clinical significance in the measurement of ionized calcium in urine.

MAGNESIUM

Physiology

Magnesium is an essential activator of >300 enzymes, including those important in glycolysis, gene replication, transcellular ion transport, muscle contraction, and oxidative phosphorylation. Numerous articles describing the clinical problems of hypomagnesemia have been reported. As examples, these reports describe

- The effects of magnesium on myocardial function and blood pressure *(35)*.
- The role of magnesium as a calcium-channel blocking agent *(36)*.
- The clinical consequences of hypomagnesemia *(37)*.
- The implications of magnesium depletion during open heart surgery *(38)* and critical care *(39)*.

TABLE 2-3. Reference Ranges for Calcium

Total calcium	
Child	2.20–2.68 mmol/L (8.8–10.7 mg/dL)
Adult	2.10–2.55 mmol/L (8.4–10.2 mg/dL)
Ionized calcium	
At birth	1.30–1.60 mmol/L (5.2–6.4 mg/dL)
Neonate	1.20–1.48 mmol/L (4.8–5.9 mg/dL)
Child	1.20–1.38 mmol/L (4.8–5.5 mg/dL)
Adult	1.16–1.32 mmol/L (4.6–5.3 mg/dL)

Regulation in the Blood

The recommended dietary intake of magnesium is 10–15 mmol/d. Rich sources of magnesium include green vegetables, meat, grains, and seafood. The small intestine absorbs from 20% to 60% of the dietary magnesium by both active and passive transport mechanisms, depending on the need. Secretions from the lower GI tract are rich in magnesium, while secretions of the upper GI tract contain relatively little magnesium. This explains why prolonged diarrhea often causes magnesium depletion, whereas vomiting has little effect.

The overall regulation of blood magnesium is controlled largely by the kidney, which avidly reabsorbs magnesium in deficiency states or readily excretes excess magnesium in overload states. The proximal tubule and the thick ascending limb of the loop of Henle are major renal regulatory sites, where ~65% of filtered magnesium may be reabsorbed. The renal threshold for magnesium is ~0.60–0.85 mmol/L (~1.46–2.07 mg/dL). Because this is close to the normal serum concentration, the kidneys rapidly excrete even slight excesses of magnesium in blood.

Although PTH has a role in regulating both magnesium and calcium, no hormonal mechanism for specifically regulating serum magnesium has yet been described. Similar to its effect on calcium, PTH increases renal reabsorption of magnesium and enhances absorption of magnesium in the intestine. Paradoxically, hypomagnesemia can depress secretion of PTH, a mechanism by which hypomagnesemia causes hypocalcemia. Although a number of other hormones affect magnesium absorption by the kidney, the effects are usually small and may be indirect. Magnesium regulation may depend on a complex interdependence of renal excretion, GI absorption, and bone exchange, and the ionized intracellular Mg concentration may be the regulatory signal *(40)*. In addition to PTH, insulin, aldosterone, and vitamin D may play a role in magnesium regulation *(41,42)*.

It is clear that the parathyroid gland is far more sensitive to a decrease in ionized calcium than magnesium *(41,43)*. As shown in Figure 2-6, a 0.065 mmol/L decrease in blood ionized calcium (about 5%) increased the PTH concentration about fourfold, while a 0.03 mmol/L decrease in ultrafiltrable magnesium (also about 5%) produced no detectable PTH response, as shown in Figure 2-7.

Regulation of magnesium is not as well characterized as that of calcium. PTH increases the release of magnesium from bone, increases the renal reabsorption of magnesium, and, with vitamin D, enhances the absorption of magnesium in the intestine. Chronic or severe acute hypomagnesemia can depress secretion of PTH, a mechanism by which hypomagnesemia causes hypocalcemia. Aldosterone apparently inhibits the renal reabsorption of magnesium, an effect opposite that of PTH.

Insulin may have both hyper- and hypomagnesemic effects. It increases magnesium concentrations in blood by increasing both intestinal and renal absorption of magnesium. However, it also increases intracellular magnesium concentrations.

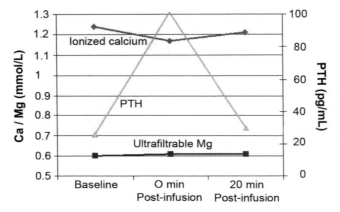

No-calcium fluid: Ion Ca, 0 mmol/L; total Mg, 1.20 mmol/L; albumin, 40 g/L; Na, 140 mmol/L; K, 5.0 mmol/L; Cl, 90 mmol/L; gluconate, 23; acetate, 27; pH, 7.40 @ 37 °C.

FIGURE 2-6. Changes in ionized calcium, PTH, and ultrafiltrable magnesium in healthy blood donors infused with a no-calcium fluid. The composition of the fluid is indicated above.

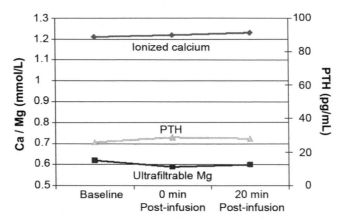

No-magnesium fluid: Ion Ca, 1.25 mmol/L; total Mg, 0 mmol/L; albumin, 40 g/L; Na, 140 mmol/L; K, 5.0 mmol/L; Cl, 90 mmol/L; gluconate, 23; acetate, 27; pH, 7.40 @ 37 °C.

FIGURE 2-7. Changes in ionized calcium, PTH, and ultrafiltrable magnesium in healthy blood donors infused with a no-magnesium fluid. The composition of the fluid is indicated above.

Distribution in Cells and Blood

The human body contains ~1 mol (24 g) of magnesium, with ~53% in the skeleton and ~46% in soft tissues such as skeletal muscle, liver, and myocardium. Magnesium is primarily an intracellular ion, with only ~1% in

Calcium, Magnesium, and Phosphate

blood and ECFs *(44)*. In a manner similar to calcium, magnesium in serum exists as protein-bound (24%), complex-bound (10%), and ionized (66%) forms (data derived from ref. *43*). As with calcium, the pH of blood apparently affects magnesium binding by proteins in the blood *(10)*.

Evaluation of Magnesium Status

Magnesium should be measured during the initial examination of ill patients with poor food intake, malabsorption disorders, hypokalemia or hypocalcemia, as well as those taking magnesium-depleting agents, such as diuretics, alcohol, or aminoglycosides. A patient with hypomagnesemia may show a variety of nonspecific symptoms, such as weaknesss, muscle cramping, and rapid heartbeat *(45)*. Although measurement of total magnesium concentrations in serum remains the usual diagnostic test for detection of magnesium abnormalities, it has limitations:

- Twenty-five to 30% of magnesium is protein-bound. Therefore, as with total and ionized calcium, total magnesium may not reflect the physiologically active free ionized magnesium.
- Magnesium is primarily an intracellular ion. Therefore, serum concentrations will not necessarily reflect the intracellular status of magnesium. Even when tissue and cellular magnesium is depleted by as much as 20%, serum magnesium concentrations may remain normal.

The magnesium load test may be the definitive method for detecting body depletion of magnesium, as was shown in 13 patients with pancreatitis and a normal concentration of magnesium in serum *(46)*. However, because a magnesium load test requires more than 48 hours to complete, its clinical use is limited. After collection of a baseline 24-hour urine, 30 mmol (729 mg) of magnesium in a 5% dextrose solution is administered intravenously (iv) as another 24-hour urine is collected. Although individuals with adequate body stores of magnesium will excrete 60–80% of the magnesium load within 24 hours, magnesium-deficient patients will excrete <50% *(27)*.

Urinary magnesium may be useful to confirm the cause of the magnesium deficit. If it is >25–30 mg/d, renal loss is suspected; if <20 mg/d, nonrenal causes should be suspected, such as inadequate intake or GI loss.

In addition to renal loss, acute hypomagnesemia can result from intracellular shifts of magnesium after the administration of glucose or amino acids *(45)*. This effect is pronounced after starvation or insulin treatment for hyperglycemia.

Causes of Hypomagnesemia

Hypomagnesemia, defined as a serum magnesium of <0.75 mmol/L (1.8 mg/dL) *(47)*, reportedly occurs in 5–50% of hospitalized patients. The common causes of hypomagnesemia are shown in Table 2-4.

TABLE 2-4. Causes of Hypomagnesemia

- Drugs such as cyclosporin, cisplatin, or diuretics
- Diabetes
- Dietary deficiency
- Alcoholism
- GI loss: diarrhea, malabsorption syndromes
- Cellular hypoxia
- Toxemia or eclampsia of pregnancy
- Loss through skin (burns)
- PTH deficit (sepsis or hypoparathyroidism)

Critical illness

Magnesium deficiency is found in a large percentage of critically ill patients *(48)*, and the presence of hypomagnesemia on admission is associated with an increased mortality rate *(49)*. In acutely ill patients, hypomagnesemia is associated with poorly controlled diabetes, coronary artery bypass surgery, alcoholism, diarrhea or GI malabsorption, malignancy, chronic obstructive pulmonary disease, and renal loss enhanced by drugs *(40,50)*. Although magnesium deficiency is more common in critically ill patients, hypermagnesemia is more often associated with a poor outcome than is hypomagnesemia *(51)*. However, hypermagnesemia may be a secondary effect of renal disease that causes the higher mortality.

Magnesium replacement therapy may be warranted if serum Mg is <0.5 mmol/L (<1.2 mg/dL) *(52)*. In acute-care patients, diagnoses commonly associated with hypomagnesemia include coronary artery disease and coronary bypass surgery, malignancy, chronic obstructive pulmonary disease, and alcoholism. Among chronic diseases, alcoholism, liver disease, and carcinoma were commonly associated with hypomagnesemia *(50)*.

Magnesium was the most common electrolyte abnormality found among pediatric intensive care unit patients *(53)*. There is also evidence that ionized magnesium may be decreased in many critically ill pediatric patients who have normal total magnesium concentrations *(54)*. In this study, the use of albumin-corrected total magnesium was not reliable in estimating the ionized magnesium.

In comparing total and ionized magnesium measurements in critical care, total hypomagnesemia is at least as common as ionized hypomagnesemia *(51,55–57)*. Interestingly, some of these studies concluded that either measurement may be used to follow magnesium status *(55,56)*, while others concluded that the ionized magnesium was more specific *(51,57)*. Another study concluded that red blood cell magnesium was the best parameter to measure because it gave a higher incidence of hypomagnesemia (37%) than did either ionized magnesium (22%) or total magnesium (16%) *(58)*. Administration of 1g of $MgSO_4$ solution to critically ill patients increased total Mg by 0.11 mmol/L and ionized Mg by 0.05 mmol/L *(59)*.

During abdominal surgery without massive transfusions, both total and ionized hypmagnesemia were common and changes in ionized and total serum Mg concentrations correlated closely *(55)*. Even though total serum Mg slightly overestimated the prevalence of hypomagnesemia, it adequately screened for hypomagnesemia *(55)*. Magnesium supplementation during cardiopulmonary bypass appeared to significantly benefit patients with hypomagnesemia by preventing ventricular tachycardia *(39,40)*.

In a review of 253 patients admitted to the ED, mild, moderate, and severe hypomagnesemia were found in 19.5%, 9.1%, and 2.5% of the patients, although these levels did not relate to mortality *(60)*.

Cardiac disorders

With magnesium ion important in energy metabolism, calcium channel regulation, and myocardial contraction and other cardiac functions, the heart is particularly vulnerable to magnesium deficiency, which can disrupt mitochondrial function and production of ATP, and promote loss of myocardial potassium. These conditions can contribute to coronary vasospasm, arrhythmias, fibrillation, infarction, and sudden death *(41)*. A low magnesium concentration enhances the potency of vasoconstrictive agents, which can produce sustained constriction of arterioles and venules. Furthermore, a low serum magnesium promotes endothelial cell dysfunction which inhibits nitric oxide (NO) release and promotes a pro-inflammatory, pro-thrombotic, and pro-atherogenic environment *(38,61)*.

Table 2-5 summarizes the cardiac disorders known to be associated with loss of myocardial and/or serum magnesium *(35)*.

Magnesium affects vascular tone by modulating the vasoconstrictive effects of hormones such as norepinephrine and angiotensin II: A high ratio of magnesium to calcium concentration antagonizes their effects, whereas a low magnesium-to-calcium ratio enhances their activity *(62)*. Calcium and magnesium ions compete for binding to the contractile proteins, with calcium initiating contraction (vasoconstriction) and magnesium inhibiting contraction (vasodilation).

TABLE 2-5. Myocardial Ischemic Syndromes and Cardiac Disorders Associated with Loss of Myocardial or Serum Magnesium

- Unstable angina
- Sudden death ischemic heart disease
- Myocardial infarction
- Cardiac arrhythmias
- Atrial fibrillation
- Congestive heart failure
- Alcoholic cardiomyopathy
- Coronary vasospasm
- Type A behavior pattern

Postoperative hypomagnesemia is common following cardiac operations and is associated with significant morbidity from arrhythmias. Hypomagnesemia also impairs the release of NO from coronary endothelium, which promotes vasoconstriction and coronary thrombosis in the early postoperative period *(38)*. To avoid this, two to four grams of magnesium sulfate in solution may be infused during or just before coming off cardiopulmonary bypass *(38)*. In patients undergoing surgery with cardiopulmonary bypass, plasma ionized (but not total) magnesium was decreased significantly by 24 hours after bypass *(63)*. By correcting serum magnesium levels during cardiopulmonary bypass, the incidence of ventricular tachycardia was reduced from 30% to 7% *(39)*.

In a study of pediatric patients undergoing surgery for congenital heart defects, magnesium supplementation was so effective in preventing ectopic tachycardia (none in 13 patients) compared to the placebo group (4 in 15), that the study was terminated after 28 patients *(64)*.

Magnesium ion acts as a calcium-channel blocker by affecting the influx of Ca ions at specific sites in the vascular membrane. In a healthy arterial cell with an adequate supply of magnesium, the gates are closed, which severely restricts the entry of Ca ions. Magnesium deficiency promotes accumulation of intracellular Ca and Na ions, leading to a state of greater contractility.

In cardiovascular disease, myocardial hypoxia accompanied by cellular magnesium deficit will more rapidly deplete ATP, leading to disruption of mitochondrial function and structure. Rapid losses of myocardial magnesium and potassium occur in both congestive heart failure and myocardial ischemic syndromes, such as sudden death, acute myocardial infarction, unstable angina, and ventricular fibrillation.

The benefit of administering magnesium in myocardial infarction is controversial. The LIMIT-2 Study on more than 2,000 patients found a distinct benefit of administering magnesium early to patients with suspected AMI *(65)*, and a rationale for the benefit of magnesium was presented *(66)*. A key factor was thought to be the timing of magnesium supplementation, which might be required to achieve benefits in high-risk patients, before reperfusion occurs. However, the very large MAGIC trial did not find a benefit of early administration of magnesium *(67)*.

In congestive heart failure, cardiac output is inadequate for the metabolic needs of tissues. Because deficiencies of magnesium and potassium lead to high blood pressure, causing increased vascular resistance, the problems of diminished cardiac output in congestive heart failure are further intensified.

Drugs

Several drugs, including diuretics, gentamicin and other aminoglycoside antibiotics, cisplatin, and cyclosporine increase renal loss of magnesium and frequently result in hypomagnesemia. The loop diuretics, such as furosemide, are especially potent in increasing renal loss of magnesium, while thiazide diuretics usually require chronic use to produce hypomagnesemia *(40,41)*.

Gentamicin inhibits reabsorption of magnesium in the renal tubule. Because hypomagnesemia intensifies the toxic side effects of digoxin, hypomagnesemia should be avoided in patients on digoxin. Cisplatin, an antineoplastic agent, is nephrotoxic and can cause profound hypomagnesemia and hyperkalemia. Good urine flow should be maintained during cisplatin therapy to maintain renal function. Cyclosporine is nephrotoxic, hepatotoxic, and hypertensive. This drug also severely inhibits renal tubular reabsorption of magnesium *(68)*. The effects of these and other drugs that cause hypomagnesemia are presented in a review *(69)*.

Diabetes

Hypomagnesemia is common in patients with both type 1 and type 2 diabetes *(42,70)*. The mechanism of magnesium deficiency appears to be magnesium loss secondary to ketoacidosis and glycosuria, along with possible abnormal intracellular-extracellular distributions of magnesium caused by insulin and other hormones such as PTH. Hypomagnesemia may lead to insulin resistance, and intensify complications frequently associated with diabetes: retinopathy, hypertension, cardiovascular disease, and increased platelet activity and thrombosis. Low serum Mg is a strong, independent risk factor for type 2 diabetes *(70)* and for insulin resistance in obese children *(71)*, and daily magnesium supplementation prevents development of insulin resistance and improves metabolic control in type 2 diabetes *(42)*. In a study of 128 patients with chronic renal failure, those with diabetes had lower total and ionized magnesium (by about 10%) than the nondiabetic renal patients *(72)*. Furthermore, another report concluded that a low serum Mg was associated with a more rapid decline of renal function in patients with type 2 diabetes *(73)*.

Magnesium deficiency appears related to increased urinary loss of magnesium that occurs with osmotic diuresis from glycosuria and with hormonal imbalances, such as decreased PTH and altered vitamin D metabolism that are related to hypomagnesemia. Hypomagnesemia may lead to insulin resistance, retinopathy, hypertension, cardiovascular disease, and increased platelet activity and thrombosis.

The American Diabetes Association issued a statement about magnesium and diabetes in 1992 *(74)*. These were among the conclusions:

1. Hypomagnesemia often results from glycosuria.
2. Hypomagnesemia probably does not play a primary role in the pathogenesis of diabetes.
3. A strong relationship exists between hypomagnesemia and insulin resistance.
4. Magnesium supplements should be given to all patients with documented hypomagnesemia, but not to all patients with diabetes.
5. Serum magnesium should be measured in those having conditions associated with magnesium deficiency, such as
   - Congestive heart failure
   - Ketoacidosis
   - K deficiency

- Diuretic use
- Pregnancy

A recent review discussed the evidence both for and against the use of magnesium supplementation in improving metabolic control in diabetes *(75)*.

Dietary deficiency

In cases of dietary magnesium deprivation, laboratory and clinical evidence of magnesium depletion become apparent after ~1 week and 6 weeks (respectively). Dietary magnesium supplementation is continually necessary because the GI tract cannot increase magnesium absorption during magnesium deprivation.

Alcoholism

Chronic alcoholism has long been associated with hypomagnesemia *(76,77)*. Both acute and chronic alcohol consumption cause renal magnesium excretion and loss of muscle magnesium. Therefore, hypomagnesemia in alcoholic patients apparently results from a combination of dietary magnesium deficiency, ketosis, vomiting, diarrhea, and hyperaldosteronism. Dextrose infusion may induce insulin secretion, which shifts magnesium back into cells. A report found that both total and ionized magnesium increased after 3 weeks of abstinence from alcohol *(77)*.

Cellular hypoxia

Cellular hypoxia leads to depletion of ATP, which causes cellular loss of magnesium. In chronic hypoxic conditions, such as decreased cardiac output, hypomagnesemia can result.

Eclampsia of pregnancy

Because magnesium requirements increase during pregnancy, hypomagnesemia may develop if intake is not also increased. Premature labor and preeclampsia or eclampsia are clearly associated with the development of hypomagnesemia. Treatment with magnesium salts is a long-accepted practice for these conditions, although extreme hypermagnesemia sometimes results from excessive doses. The fetus may also be affected by hypermagnesemia in such cases.

Diarrhea

Because secretions from the lower GI tract are relatively rich in magnesium, diarrhea, malabsorption syndromes, bowel resection, etc., are common causes of magnesium depletion.

Burns

Loss via burns is associated with a general loss of electrolytes.

PTH deficiency

Because PTH is one of the hormones that appear to increase tubular reabsorption of Mg in the kidney, a deficit of PTH, such as in primary hypoparathyroidism, will lead to hypomagnesemia. Sepsis inhibits PTH secretion and is also associated with hypomagnesemia.

Other diseases

Magnesium deficiency may promote the development and progression of renal stone formation and other renal calcification *(78,79)*. Magnesium ion may prevent calcification by chelating anions that would otherwise form insoluble salts with calcium.

Patients with Paget's disease who are hypomagnesemic tend to have more active disease. Because urinary magnesium is not significantly increased in Paget's disease, magnesium deficiency may by caused by increased uptake of magnesium into bone. For these reasons, patients with Paget's disease should increase their intake of magnesium to ~10 mmol/d (240 mg/d) *(80)*.

Treatment of Hypomagnesemia

Magnesium depletion should be prevented, if possible, by administering magnesium supplements. However, if magnesium infusion is necessary, here is a guideline: 25 mmol of magnesium sulfate solution given intravenously over 12 hours increased serum magnesium concentration to about 1.5 mmol/L, while 50 mmol per 12 hours increased serum magnesium to about 2 mmol/L *(81)*. Renal function should be assessed to determine the level of magnesium supplementation and prevent hypermagnesemia *(81)*. In another study, administration of 1g of MgSO4 in solution to critically ill patients increased total magnesium by 0.11 mmol/L and ionized magnesium by 0.05 mmol/L *(59)*. When drugs that deplete magnesium, such as gentamicin, diuretics, cisplatin, and cyclosporine, are given, magnesium sulfate solution can be given intravenously once or more daily, as guided by serum magnesium measurements.

Women with toxemia of pregnancy (eclampsia) are sometimes given very large doses of magnesium sulfate intravenously (175 mmol/24h), which can push serum magnesium to potentially dangerous levels of 4.0 mmol/L and above *(81)*.

Causes of Hypermagnesemia

Because the kidney can readily excrete magnesium, spontaneous hypermagnesemia is rarely observed in patients with normal renal function. Hypermagnesemia may result from administration of magnesium-containing antacids, enemas, or parenteral nutrition to patients with renal insufficiency *(82)*. Hypermagnesemia is sometimes seen in adrenal insufficiency, probably related to diminished aldosterone. The most common cause of increased plasma magnesium without chronic renal failure is found in treatment of eclampsia with magnesium salts during pregnancy.

Proper Collection and Handling of Samples

If possible, the patient should fast before blood collection. Because of the threefold-higher concentration of magnesium inside erythrocytes, serum should be separated or isolated (serum separator gel) from the clot as soon as possible, and hemolyzed samples are generally not acceptable. Samples that contain the anticoagulants citrate, oxalate, or ethylenediaminetetraacetic acid (EDTA) should not be used because they chelate magnesium tightly and interfere with methods (including atomic absorption to some degree) that are used to measure magnesium (and calcium).

Preanalytical variables for ionized magnesium are similar to those for ionized calcium, which are affected by sample pH and anticoagulants or other additives in the blood collection tube. The pH affects ionized Mg inversely: for each 0.1 pH unit change, the ionized Mg changes by about 0.01 mmol/L *(10)*. Ordinary sodium heparin decreases ionized magnesium by chelation, with ionized magnesium decreasing by about 0.01 mmol/L for each 25 IU/mL of heparin *(83)*. Balanced heparins do not have this effect. Citrate lowers the ionized magnesium concentration markedly *(84)*, and silicone found in some, but not all, blood collection devices apparently increases the ionized magnesium result *(83)*.

Diagnosis of Magnesium Deficiency

While measurement of total magnesium concentrations in serum remains the usual diagnostic test for the detection of magnesium abnormalities, total magnesium concentrations have two limitations. First, about 30% of magnesium is protein bound. Therefore, as with total vs. ionized calcium, total magnesium may not reflect the physiologically active ionized magnesium. Second, and probably of greater significance, because magnesium is primarily an intracellular ion, concentrations of either total or ionized magnesium in blood will not necessarily reflect the presumably more relevant intracellular magnesium. Symptomatic hypomagnesemia may not appear until serum total magnesium levels fall below 0.5 mmol/L (well below the typical reference limit), which makes associations between serum magnesium levels and benefit of treatment difficult *(40)*. Clearly, a more sensitive means of detecting subtle magnesium deficiencies would be helpful.

A *Magnesium Retention Test* may detect body depletion of magnesium in patients with a normal concentration in serum. However, a magnesium retention test requires more than 48 hours to complete. After collecting a baseline 24-hour urine, 30 mmol of $MgSO_4$ is administered intravenously over 12 hours while another 24-hour urine is collected. Individuals with adequate body stores of magnesium will excrete 60% (18 mmol) or more of the magnesium load within 24 hours, while magnesium-deficient patients excrete less than 50% (15 mmol) *(40)*.

Although less analytically reliable than measurement of ionized calcium, the routine measurement of ionized magnesium is now possible with the development of magnesium ion-selective electrodes. The sensors for these electrodes were developed from studies over many years on more than 200 ionophores produced by the late Dr. Wilhelm Simon and his colleagues. However, all Mg ion-sensitive electrodes are also sensitive to calcium ions. While this interference by calcium ion is corrected by simultaneous measurement of the calcium ion concentration, this additional test adds to the variability of the ionized magnesium measurement *(85)*. While some studies conclude that measurements of ionized magnesium improve clinical accuracy *(51,57)*, others find that either total or ionized magnesium provides about the same clinical value in detecting hypomagnesemia *(39,55,56)*.

Reference Ranges for Magnesium

Reference ranges for total and ionized magnesium are shown in Table 2-6 *(86)*. Concentrations of magnesium in serum are slightly higher in older children and adults.

Based on a lower reference limit for serum magnesium of 0.66 mmol/L, the incidence of hypomagnesemia in hospitalized patients has been reported to be about 5–50%, with differences in reference ranges, methods, and populations studied accounting for some of this variation. One report suggests that this standard lower limit will frequently miss hypomagnesemia and suggests that increasing the lower reference limit to 0.70–0.80 mmol/L would miss fewer cases of hypomagnesemia *(87)*. However, because this would also increase the rate of false positive diagnosis of hypomagnesemia, an actual clinical study would need to be conducted.

TABLE 2-6. Reference Ranges for Magnesium

Total serum magnesium	
Newborns	0.50–0.90 mmol/L (1.22–2.19 mg/dL)
Adults	0.65–1.05 mmol/L (1.58–2.55 mg/dL)
Ionized blood magnesium	0.40–0.62 mmol/L (0.97–1.52 mg/dL)
Erythrocytes	1.65–2.65 mmol/L (4.01–6.44 mg/dL)
Cerebrospinal fluid	1.0–1.40 mmol/L (2.43–3.40 mg/dL)
Urine	1–5 mmol/d

PHOSPHATE

Physiology

Phosphate compounds are in all cells and participate in numerous biochemical processes. Phosphate is a component of DNA and RNA, phospholipids, high-energy compounds such as ATP and creatine phosphate, and many coenzymes. Phosphate is required for glucose to enter cells; in fact, glycolysis is stimulated by high concentrations of intracellular inorganic phosphorus and inhibited by low concentrations. Furthermore, insulin resistance may be a consequence of hypophosphatemia *(88)*.

When both phosphate depletion and hypophosphatemia are present, serious biochemical abnormalities often result: impaired myocardial function, respiratory muscle paralysis, central nervous system disorders, and skeletal muscle changes *(88)*. These may result from hypophosphatemia inhibiting mitochondrial respiration and ATP synthesis, processes that are vital to cellular survival.

Regulation in Blood

Although cellular shifts of phosphate can affect phosphate concentrations in the blood and ECF, absorption from the intestine and excretion by the kidney are the dominant homeostatic mechanisms. Because absorption by the intestine fluctuates widely, the kidney is responsible for precise regulation of phosphate concentrations in the blood.

The renal tubules normally reabsorb >90% of phosphate filtered at the glomerulus *(82)*. However, PTH can have a dramatic effect on phosphate regulation by inhibiting the normal tubular reabsorption of phosphate, increasing loss in the urine, and decreasing serum phosphate concentrations in blood.

Vitamin D increases phosphate concentrations in the blood by promoting phosphate absorption in the intestine and phosphate reabsorption in the kidney. In fact, phosphate may have a direct effect on parathyroid cells to enhance PTH secretion *(89)*.

By decreasing renal excretion of phosphate, growth hormone increases concentrations of phosphate in the blood *(90)*. A number of peptides, collectively referred to as "phosphatonins," have been recently identified as playing a role in the development of both hypophosphatemic and hyperphosphatemic disorders *(91)*. In addition, there are several other hormones, dietary factors, and drugs that may affect phosphate absorption and excretion.

Distribution in Cells and Blood

About 80% of phosphate is in bone, mostly in the form of hydroxyapatite $[Ca_{10}(PO_4)_6(OH)_2]$. Phosphate in blood is either absorbed from dietary sources or resorbed from bone. Most phosphate is found within cells, and the transport of glucose into cells is accompanied by an influx of phosphate.

Calcium, Magnesium, and Phosphate

The total concentration of phosphate (expressed as phosphorus) in blood is ~3.87 mmol/L (~12 mg/dL), mostly as organic phosphate. Only ~0.97–1.29 mmol/L (~3–4 mg/dL) is inorganic phosphate, with 10–15% bound to anions and 90% circulating as the free ion *(82)*. At pH 7.4, inorganic phosphate is about 75% HPO_4^{2-} and 25% $H_2PO_4^{-}$.

Evaluation of Hypophosphatemia

Moderate hypophosphatemia is diagnosed at ~0.48–0.81 mmol/L (~1.5–2.5 mg/dL), whereas a phosphate concentration <0.48 mmol/L (<1.5 mg/dL) is considered severe hypophosphatemia *(82)*.

When hypophosphatemia has been confirmed, it is important to determine whether the cause is renal loss, GI loss, or transcellular shifts. The urinary phosphate excretion should be evaluated next, as shown in Figure 2-8:

- A low urinary phosphate excretion [<3.3 mmol (or 100 mg)/d)] suggests a nonrenal cause: a transcellular shift (such as with insulin administration), a GI loss, or a dietary deficiency of phosphate. In severe phosphate depletion, phosphate is nearly absent from the urine.
- A high urinary phosphate excretion [>3.3 mmol (or 100 mg)/d)] indicates a renal loss of phosphate caused by a renal tubular defect, use of diuretics, hyperparathyroidism, or a vitamin D defect or deficiency.

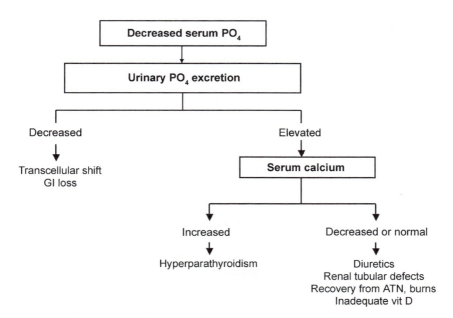

FIGURE 2-8. Use of laboratory tests in evaluation of hypophosphatemia. vit D, vitamin D; ATN, acute tubular necrosis.

Causes of Hypophosphatemia

The incidence of hypophosphatemia in hospitalized patients varies greatly, depending on the population studied *(88,92)*. For example, hypophosphatemia may occur in 1% of general patients upon admission. The percentage increases with length of stay, critical illness, and alcoholism.

Hypophosphatemia is commonly caused by transcellular shifts of phosphate, GI malabsorption, or renal loss. Although dietary deficiency rarely causes hypophosphatemia, massive phosphate depletion may be seen during major societal upheavals such as those that occurred during World War II, when fertilizers were not available *(88)*.

The mechanisms of hypophosphatemia are described below:

Transcellular shifts

Because the movement of glucose into cells is accompanied by phosphate, the administration of glucose or insulin will lead to an influx of phosphate into cells that may in turn cause hypophosphatemia. By a complex mechanism involving glycolysis and formation of phosphorylated intermediates in cells, respiratory alkalosis also substantially enhances phosphate uptake by cells *(88,92)*. Urinary phosphate is usually decreased.

GI losses

As with many electrolyte disorders, diarrhea and vomiting can diminish GI absorption of phosphate. Because the divalent cations in antacids such as aluminum hydroxide, magnesium hydroxide, or aluminum carbonate bind phosphate and prevent its absorption in the gut, hypophosphatemia often results from chronic excessive use of such antacids *(93)*. Urine phosphate is usually low.

Renal losses

Causes of renal loss include primary hyperparathyroidism, diuretics, hypomagnesemia, or defects in renal tubular absorption of phosphate. Urine phosphate is normal or elevated.

Mixed causes

Hypophosphatemia can be caused by a mix of factors *(82,88)*:

- In diabetic ketoacidosis, the combined effects of acidosis, glycosuria, ketonuria, and insulin therapy deplete phosphate, both by renal loss and cellular uptake. Serum concentrations may decrease rapidly.
- Acidosis induces mobilization of phosphate from bone and tissues, along with increased renal loss.
- Alcoholism leads to hypophosphatemia by increasing renal loss and decreasing GI absorption.
- During recovery from thermal burns, phosphate is shifted into cells and secreted into renal tubules, where it is lost in urine.

Symptoms of Hypophosphatemia

Severe hypophosphatemia usually results in decreased concentrations of phosphate-containing compounds, such as ATP and membrane phospholipids. These deficiencies are responsible for symptoms of hypophosphatemia, such as muscle weakness, respiratory and myocardial insufficiency, and hepatocellular damage *(82,88)*.

Hypophosphatemic patients have a marked increase in urinary calcium and magnesium excretion, caused by both increased bone loss and altered renal tubular handling of these ions. Phosphate depletion also suppresses PTH secretion, which leads to further urinary loss of these ions.

Moderate hypophosphatemia can lead to weakness of pulmonary muscles *(82,88)* and can prolong weaning of patients from a ventilator. These effects are often reversed by phosphate repletion.

More severe hypophosphatemia is associated with life-threatening seizures, coma, and dysfunction of respiratory and myocardial muscles. Other undesirable cardiovascular effects include impaired mitochondrial oxygen consumption and decreased sensitivity to inotropic agents, such as epinephrine *(88)*.

Phosphate depletion reduces 2,3-DPG in erythrocytes, which increases the affinity of Hb for oxygen. This can reduce oxygen release to tissues by 10–15% in severe hypophosphatemia *(88)*. If hypophospatemia occurs suddenly, it can cause hemolysis *(94)*.

Phosphate depletion causes a rapid breakdown of bone matrix, even more so than that caused by either severe hypocalcemia or vitamin D depletion. This effects appear to be independent of vitamin D or PTH *(88)*.

Treatment of Hypophosphatemia

Mild hypophosphatemia is common in hospitalized patients and is often not treated or may be treated with oral phosphate supplements such as milk *(82)*. Phosphorus supplements should be avoided in patients with renal failure or with parathyroid disorders *(93)*. In cases of severe hypophosphatemia (<1.0 mg/dL or 0.3 mmol/L), solutions of phosphate may be given intravenously. Persistent hypophosphatemia is associated with a poor prognosis [mortality rate 20% *(88)*]. Studies have shown that treatment of hypophosphatemia may reverse not only hypophosphatemia, but other electrolyte and acid-base disorders as well. The decision to treat should be based on factors such as severity, duration, and presence of symptoms *(82,88,89,94)*.

Evaluation of Hyperphosphatemia

Because hyperphosphatemia is often associated with common types of metabolic acidosis (ketoacidosis and lactate acidosis), and either acute or chronic renal failure, hyperphosphatemia is relatively common in hospitalized populations. An elevated serum phosphorus should be followed up with measurements of electrolytes, calcium, creatinine, and magnesium *(95)*.

A review of several texts on phosphate does not give specific ranges for mild, moderate, or severe hyperphosphatemia *(27,88,96)*. With an adult reference range of 0.87–1.45 mmol/L (2.7–4.5 mg/dL), a general range for mild to moderate hyperphosphatemia may be 1.6–2.9 mmol/L (5–9 mg/dL), with severe hyperphosphatemia >2.9 mmol/L (9 mg/dL).

As shown in Figure 2-9, further investigation of an elevated serum phosphate should be followed by evaluation of renal function:

- An elevated serum creatinine (or decreased creatinine clearance) indicates that diminished renal function is the cause of the hyperphosphatemia.
- If serum creatinine indicates normal renal function, urinary phosphate excretion should then be determined. A normal urinary phosphate suggests increased phosphate reabsorption such as in hypoparathyroidism. An increased phosphate excretion >48 mmol (or 1500 mg)/d suggests an increased phosphate load to the kidneys such as by increased dietary intake, increased cellular breakdown (chemotherapy, rhabdomyolysis), or excess administration of phosphate by intravenous or intestinal routes *(96)*.

Causes of Hyperphosphatemia

Hyperphosphatemia usually results from one of these mechanisms:

- Renal failure, which reduces the renal excretion of phosphate. This is by far the most common condition associated with hyperphosphatemia and is often a contributing factor with excessive intake of phosphate-containing antacids, laxatives, enemas, vitamin D, or milk.
- Increased transfer of phosphate from intracellular stores into (extracellular) plasma. This occurs in both lactate and ketoacidosis (note that these often eventually result in hypophosphatemia), tissue damage from

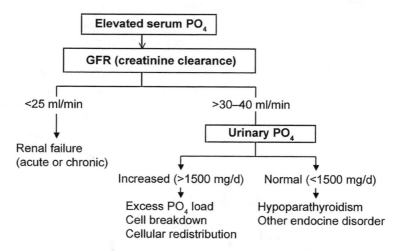

FIGURE 2-9. Use of laboratory tests in evaluation of hyperphosphatemia. GFR, glomerular filtration rate.

chemotherapy for various malignancies such as leukemias or lymphomas, acute hemolysis, and rhabdomyolysis.
- Increased cellular catabolism or injury, such as by rhabdomyolysis, trauma, burns, massive hemolysis, malignant hyperthermia, leukemia and other neoplasms, and exhaustive exercise.
- Endocrine disorders should be suspected when renal function and urinary phosphate excretion are normal. These include hypoparathyroidism, PTH resistance, hyperthyroidism, or postmenopausal state *(88)*.
- Excess administration, such as excess oral intake, intravenous phosphate solutions, use of phosphate-containing enemas, or vitamin D intoxication.

Clinical Consequences of Hyperphosphatemia

Hyperphosphatemia that develops acutely (within hours or days) may be associated with these findings *(88)*:

- Hypocalcemia, possibly with secondary hyperparathyroidism.
- Precipitation of calcium and phosphate salts in various tissues (which may cause the remaining problems in this list). Excess phosphate can be taken up into vascular smooth muscle cells by a sodium-phosphate cotransporter, with deposition of calcium in these cells that leads to calcification and arteriosclerosis *(95)*.
- Decreased glomerular filtration rate and oliguria within the renal tubule.
- Conductance disturbances in the heart, leading to arrhythmias and hypotension.
- Impaired O_2 diffusion in the alveoli.
- Skin and vision problems.

Treatment of Hyperphosphatemia

Treatment should be directed at eliminating the source of phosphate, removing the excess phosphate, and correcting any associated electrolyte disorder such as hypocalcemia, hypomagnesemia, or hypokalemia *(82,96)*. Management depends on whether the hyperphosphatemia has occurred acutely or chronically. Acute hyperphosphatemia is serious and may require insulin and dextrose, which transfer phosphate into cells, or acetazolamide, which increases urinary excretion, to rapidly normalize blood levels. If renal failure is present in serious cases of hyperphosphatemia, peritoneal dialysis or hemodialysis may be necessary *(82,96)*.

Chronic hyperphospatemia may be treated by reducing intestinal absorption, such as by ingesting a phosphate-binding salt. However, if renal failure is present, treatment with a calcium salt may be preferable *(82)*. Dietary intake of phosphate should be kept at <6.5 mmol (200 mg)/d.

Reference Ranges for Phosphate

Phosphate levels are normally higher in children and postmenopausal women. Reference ranges for phosphate are shown in Table 2-7 *(89,97)*.

TABLE 2-7. Reference Ranges for Phosphate

Neonate	1.13–2.78 mmol/L (3.5–8.6 mg/dL)
Child	1.45–1.78 mmol/L (4.5–5.5 mg/dL)
Adult	0.87–1.34 mmol/L (2.7–4.5 mg/dL)

REFERENCES

1. Ariyan CE, Sosa JA. Assessment and management of patients with abnormal calcium. Crit Care Med 2004;32:S146–S154.
2. Zaloga GP, Chernow B. Calcium metabolism. In: Geelhoed GW, Chernow B, eds. Endocrine aspects of acute illness. New York: Churchill Livingstone, 1985:169–204.
3. Toffaletti J, Nissenson R, Endres D, et al. Influence of continuous infusion of citrate on responses of immunoreactive PTH, calcium and magnesium components, and other electrolytes in normal adults during plateletapheresis. J Clin Endocrinol Metab 1985;60:874–9.
4. Saha H, Harmoinen A, Pietila K, et al. Measurement of serum ionized versus total levels of magnesium and calcium in hemodialysis patients. Clin Nephrol 1996;46:326–31.
5. Ahren B, Bergenfelz A. Effects of minor increase in serum calcium on the immunoheterogeneity of PTH in healthy subjects and in patients with primary hyperparathyroidism. Horm Res 1995;43:294–9.
6. Mundy GR, Guise TA. Hormonal control of calcium homeostasis. Clin Chem 1999;45:1347–52.
7. Felsenfeld AJ, Machado L, Rodriguez M. The relationship between serum calcitonin and calcium in the hemodialysis patient. Am J Kidney Dis 1993;21:292–9.
8. Mundy GR, Edwards JR. PTH-related peptide (PTHrP) in hypercalcemia. J Am Soc Nephrol 2008;19:672–5.
9. Toffaletti J, Gitelman HJ, Savory J. Separation and quantitation of serum constituents associated with calcium by gel filtration. Clin Chem 1976;22:1968–72.
10. Wang S, McDonnell B, Sedor FS, et al. pH effects on measurements of ionized calcium and ionized magnesium in blood. Clin Chim Acta 2002;126:947–50.
11. Gray TA, Patterson CR. The clinical value of ionized calcium asays. Ann Clin Biochem 1988;25:210–9.
12. Sokoll LJ, Donovan PI, Udelsman R. The National Academy of Clinical Biochemistry laboratory medicine practice guidelines for intraoperative parathyroid hormone. Point of Care 2007;6:253–60.
13. Eleraj DM, Remaley AT, Simonds WF, et al. Utility of rapid intraoperative parathyroid hormone assay to predict severe postoperative hypocalcemia after reoperation for hyperparathyroidism. Surgery 2002;132:1028–34.
14. Ladenson JM, Lewis JW, Boyd JC. Failure of total calcium corrected for protein, albumin, and pH to correctly assess free calcium status. J Clin Endocrinol Metab 1978;46:986–93.
15. Byrnes MC, Huynh K, Helmer SD, et al. A comparison of corrected serum calcium levels to ionized calcium levels among critically ill surgical patients. Amer J Surgery 2005;189:310–4.
16. Jennichjen S, van der Voort PHJ, Gerritsen RT, et al. Albumin-adjusted calcium is not suitable for diagnosis of hyper- and hypocalcemia in the critically ill. Crit Care Med 2003;31:1389–93.

17. Dickerson RN, Alexander KH, Minard G, et al. Accuracy of methods to estimate ionized and "corrected" serum calcium concentrations in critically ill multiple trauma patients receiving specialized nutrition support. J Parenter Enteral Nutr 2004;28:133–41.
18. Carney SL, Gillies AHG. Effect of an optimum dialysis fluid calcium concentration on calcium mass transfer during maintenance hemodialysis. Clin Nephrol 1985;24:28–30.
19. Wandrup J. Critical analytical and clinical aspects of ionized calcium in neonates. Clin Chem 1989;35:2027–33.
20. Vincent J-L, Jankowski S. Why should ionized calcium be determined in acutely ill patients? Acta Anaesthesiol Scand 1995;39[Suppl 107]:281–6.
21. Narins RG, ed. Maxwell and Kleeman's clinical disorders of fluid and electrolyte metabolism, 5th ed. New York: McGraw-Hill, 1994:1082.
22. Desai TK, Carlson RW, Thill-Baharozian M, et al. A direct relationship between ionized calcium and arterial pressure among patients in an intensive care unit. Crit Care Med 1988;16:578–82.
23. Anast CS, Winnacker JL, Forte LF, et al. Impaired release of parathyroid hormone in magnesium deficiency. J Clin Endocrinol Metab 1976;42:707–17.
24. Medalle R, Waterhouse C, Hahn TJ. Vitamin D resistance in magnesium deficiency. Am J Clin Nutr 1976;29:854–8.
25. Lafferty FW. Differential diagnosis of hypercalcemia. J Bone Miner Res 1991;6[Suppl 2]:S51–9.
26. Health H III, Earll JM, Schaaf M, et al. Serum ionized calcium during bed rest in fracture patients and normal man. Metabolism 1972;21:633–40.
27. Zaloga GP, Chernow B. Divalent ions: calcium, magnesium, and phosphorus. In: Chernow B, ed. The pharmacologic approach to the critically ill patient, 2nd ed. Baltimore: Williams and Wilkins, 1988:621–7.
28. Ladenson JH, Lewis JW, McDonald JM, et al. Relationship of free and total calcium in hypercalcemic conditions. J Clin Endocrinol Metab 1979;48:393–7.
29. Fleetwood MK, Quinton L, Wolfe J, et al. Rapid PTH assay by simple modifications of Nichols intact PTH-parathyroid hormone assay kit. Clin Chem 1996;42:1498.
30. Triponez F, Kebebew E, Dosseh D, et al. Less-than-subtotal parathyroidectomy increases the risk of persistent/recurrent hyperparathyroidism after parathyroidectomy in tertiary hyperparathyroidism after renal transplantation. Surgery 2006; 140:997–9.
31. Lepage R, Whittom S, Bertrand S, et al. Superiority of dynamic over static reference intervals for intact, midmolecule, and C-terminal parathyrin in evaluating calcemic disorders. Clin Chem 1992;38:2129–35.
32. Toffaletti J, Abrams B. Effects of in vivo and in vitro production of lactic acid on ionized, protein-bound, and complex-bound calcium in blood. Clin Chem 1989;35:935–8.
33. Toffaletti JG. Use of novel preparations of heparin to eliminate interference in ionized calcium measurements: have all the problems been solved? [Editorial]. Clin Chem 1994;40:508–9.
34. Toffaletti JG, Wildermann RF. The effects of heparin anticoagulants and fill volume in blood gas syringes on ionized calcium and magnesium measurements. Clin Chim Acta 2001;304:147–51.
35. Gomez MN. Magnesium and cardiovascular disease. Anesthesiology 1998;89:222–40.
36. Iseri LT, French JH. Magnesium: nature's physiologic calcium blocker [Editorial]. Am Heart J 1984;108:188–93.

37. Gums JG. Clinical significance of magnesium: a review. Drug Intell Clin Pharm 1987;21:240–6.
38. Pearson PJ, Evora PRB, Seccombe JF, et al. Hypomagnesemia inhibits nitric oxide release from coronary endothelium: protective role of magnesium infusion after cardiac operations. Ann Thorac Surg 1998;65:967–72.
39. Wilkes NJ, Mallett SV, Peachey T, et al. Correction of ionized plasma magnesium during cardiopulmonary bypass reduces risk of postoperative cardiac arrhythmia. Anesth Analg 2002;95:828–34.
40. Noronha JL, Matuschak GM. Magnesium in critical illness: metabolism, assessment, and treatment. Intensive Care Med 2002;28:667–79.
41. Topf JM, Murray PT. Hypomagnesemia and hypermagnesemia. Rev Endocrin Metab Disorders 2003;4:195–206.
42. Rodriguez-Moran M, Guerrero-Romero F. Oral magnesium supplementation improves insulin sensitivity and metabolic control in type 2 diabetic subjects. Diabetes Care 2003;26:1147–52.
43. Toffaletti J, Cooper D, Lobaugh B. The response of parathyroid hormone to specific changes in either ionized calcium, ionized magnesium, or protein-bound calcium in humans. Metabolism 1991;40:814–8.
44. Elin RJ. Magnesium: the fifth but forgotten electrolyte. Am J Clin Pathol 1994;102:616–22.
45. Hypomagnesemia. www.emedicine.com/emerg/TOPIC274.HTM (Accessed Nov 2008).
46. Papazacharion IM, Martinez-Isla A, Efthimiou E, et al. Magnesium deficiency in patients with chronic pancreatitis identified by an intravenous loading test. Clin Chim Acta 2000;302:145–54.
47. White JR Jr, Campbell RK. Magnesium and diabetes: a review. Ann Pharmacother 1993;27:775–80.
48. Salem M, Munoz R, Chernow B. Hypomagnesemia in critical illness. Crit Care Clin 1991;7:225–52.
49. Rubeiz GJ, Thill-Baharozian M, Hardie D, et al. Association of hypomagnesemia and mortality in acutely ill medical patients. Crit Care Med 1993;21:203–9.
50. Lum G. Hypomagnesemia in acute and chronic care patient populations. Am J Clin Pathol 1992;97:827–30.
51. Escuela MP, Guerra M, Anon JM, et al. Total and ionized magnesium in critically ill patients. Intensive Care Med 2005;31:151–6.
52. Chernow B, Bamberger S, Stoiko M. Hypomagnesemia in patients in postoperative intensive care. Chest 1989;95:391–7.
53. Broner CW, Stidham GL, Westenkirchner DF, et al. Hypermagnesemia and hypocalcemia as predictors of high mortality in critically ill pediatric patients. Crit Care Med 1990;18:921–8.
54. Fiser RT, Torres A, Butch AW, et al. Ionized magnesium concentrations in critically ill children. Crit Care Med 1998;26:2048–52.
55. Lanzinger MJ, Moretti EW, Toffaletti JG, et al. The relationship between ionized and total serum magnesium concentrations during abdominal surgery. J Clin Anesthesia 2003;15:245–9.
56. Koch SM, Warters RD, Mehlhorn U. The simultaneous measurement of ionized and total calcium and ionized and total magnesium in intensive care unit patients. J Crit Care 2002;17:203–5.
57. Huijgen HJ, Soesan M, Sanders R, et al. Magnesium levels in critically ill patients. What should we measure? Am J Clin Pathol 2000;114:688–95.

58. Malon A, Brockman C, Fijalkowska-Morawska J, et al. Ionized magnesium in erythrocytes—the best magnesium parameter to observe hypo- or hypermagnesemia. Clin Chem Acta 2004;349:67–73.
59. Barrera R, Fleisher M, Groeger J. Ionized magnesium supplementation in critically ill patients: comparing ionized and total magnesium. J Crit Care 2000;15:36–40.
60. Stalnikowicz R. The significance of routine serum magnesium determination in the ED. Am J Emerg Med 2003;21:444–7.
61. Maier JAM, Malpuech-Brugere C, Zimowska W, et al. Low magnesium promotes endothelial cell dysfunction: implications for atherosclerosis, inflammation, and thrombosis. Biochim Biophysica Acta 2004;1689:13–21.
62. Rude R, Manoogian C, Ehrlich L, et al. Mechanism of BP regulation by Mg in man. Magnesium 1989;8:266–73.
63. Brookes CIO, Fry CH. Ionised magnesium and calcium in plasma from healthy volunteers and patients undergoing cardiopulmonary bypass. Br Heart J 1993;69:404–8.
64. Dorman BH, Sade RM, Burnette JS, et al. Magnesium supplementation in the prevention of arrhythmias in pediatric patients undergoing surgery for congenital heart defects. Am Heart J 2000;139:522–8.
65. Woods KL, Fletcher S. Long-term outcome after intravenous magnesium sulphate in suspected AMI: the second Leicester intravenous magnesium intervention trial (LIMIT-2). Lancet 1994;343:816–9.
66. Schechter M, Kaplinsky E, Rabinowitz B. The rationale of magnesium supplementation in acute myocardial infarction. A review of the literature. Arch Intern Med 1992;152:2189–96.
67. Early administration of intravenous magnesium to high-risk patients with acute myocardial infarction in the MAGIC Trial: a randomized controlled trial. Lancet 2002;360:1189–96.
68. June CH, Thompson CB, Kennedy MS, et al. Profound hypomagnesemia and renal magnesium wasting associated with the use of cyclosporine for marrow transplantation. Transplantation 1985;39:620–4.
69. Atsmon J, Dolev E. Drug-induced hypomagnesemia: scope and management. Drug Safety 2005;28:763–88.
70. Kao WHL, Folsom AR, Nieto FJ, et al. Serum and dietary magnesium and the risk for type 2 diabetes mellitus. The atherosclerosis risk in communities study. Arch Intern Med 1999;159:2151–9.
71. Huerta MG, Roemmich JN, Kington ML, et al. Magnesium deficiency is associated with insulin resistance in obese children. Diabetes Care 2005;28:1175–81.
72. Dewitte K, Dhondt A, Giri M, et al. Differences in serum ionized and total magnesium values during chronic renal failure between nondiabetic and diabetic patients. Diabetes Care 2004;27:2503–5.
73. Pham PC, Pham PM, Pham PA, et al. Lower serum magnesium levels are associated with more rapid decline of renal function in patients with diabetes mellitus type 2. Clin Nephrol 2005;63:429–36.
74. American Diabetes Association. Magnesium supplementation in the treatment of diabetes. Diabetes Care 1992;15:1065–7.
75. Sales CH, Pedrosa LDFC. Magnesium and diabetes mellitus: their relation. Clin Nutrition 2006;25:554–62.
76. Flink EB. Magnesium deficiency in alcoholism. Alcohol Clin Exp Res 1986;10:590–4.
77. Hristova EN, Rehak NN, Cecco S, et al. Serum ionized magnesium in chronic alcoholism: is it really decreased? Clin Chem 1997;43:394–9.

78. Revusova V, Zvara V, Karlikova L, et al. Prognosis of urolithiasis and nephrocalcinosis in hypomagnesemia. Czech Med 1985;8:207–13.
79. Wei M, Esbaei K, Bargman J, et al. Relationship between serum magnesium, parathyroid hormone, and vascular calcification in patients on dialysis: a literature review. Peritoneal Dialysis International 2006;26:366–73.
80. Taylor WM. Low serum magnesium concentration in Paget's disease of bone (osteitis deformans). Ann Clin Biochem 1985;22[Pt 6]:591–5.
81. Oster JR, Epstein M. Management of magnesium depletion. Am J Nephrol 1988;8:349–54.
82. Weiss-Guillet E-M, Takala J, Jakob SM. Diagnosis and management of electrolyte emergencies. Best Practice & Research Clin Endocrinol Metab 2003;17:623–51.
83. Ritter C, Ghahramani M, Marsoner HJ. More on the measurement of ionized magnesium in whole blood. Scand J Clin Lab Invest 1996;56,Suppl 224:275–80.
84. Zoppi F, de Gasperi A, Guagnellini E, et al. Measurement of ionized magnesium with AVL 988/4 electrolyte analyzer: preliminary analytical and clinical results. Scand J Clin Lab Invest 1996;56,Suppl 224:259–74.
85. Cecco SA, Hristova EN, Rehak NN, et al. Clinically important intermethod differences for physiologically abnormal ionized magnesium results. Am J Clin Pathol 1997;108:564–9.
86. Hristova EN, Cecco S, Niemela JE, Rehak NN, Elin RJ. Analyzer-dependent differences in results for ionized calcium, ionized magnesium, sodium, and pH. Clin Chem 1995;41:1649–53.
87. Leibscher D-H, Liebscher D-E. About the misdiagnosis of magnesium deficiency. J Am Coll Nutr 2004;23:730S.
88. Levine BS, Kleeman CR. Hypophosphatemia and hyperphosphatemia: clinical and pathophysiologic aspects. In: Narins RG, ed. Maxwell and Kleeman's clinical disorders of fluid and electrolyte metabolism, 5th ed. New York: McGraw-Hill, 1994:1045–97.
89. Kates DM, Sherrard DJ, Andress DL. Evidence that seum phosphate is independently associated with serum PTH in patients with chronic renal failure. Am J Kidney Dis 1997;30:809–13.
90. Slatopolsky E, Rutherford WE, Rosenbaum R, et al. Hyperphosphatemia. Clin Nephrol 1977;7:138–46.
91. Shaikh A, Berndt T, Kumar R. Regulation of phosphate homeostasis by the phosphatonins and other novel mediators. Pediatr Nephrol 2008;23:1203–10.
92. Toffaletti JG. Calcium, magnesium, and phosphate. In: McClatchey KD, ed. Clinical laboratory medicine. Baltimore: Williams and Wilkins, 1994:387–401.
93. King AL, Sica DA, Miller G, et al. Severe hypophosphatemia in a general hospital population. South Med J 1987;80;831–5.
94. Hypophosphatemia. www.ecureme.com/emyhealth/data/hypophospatemia.asp (Accessed Nov 2008).
95. Patterson LA. Hyperphosphatemia. www.emedicine.com/emerg/TOPIC266.HTM (Accessed Nov 2008).
96. Chester WL, Zaloga GP, Chernow B. Phosphate problems in the critically ill patient. In: Geelhoed GW, Chernow B, eds. Endocrine aspect of acute illness. New York: Churchill Livingstone, 1985:205–16.
97. Toffaletti JG. Electrolytes. In: Bishop ML, Duben-Engelkirk JL, Fody EP, eds. Clinical chemistry: principles, procedures, correlations, 3rd ed. Philadelphia: Lippincott, 1996:255–78.

Chapter 3

Electrolytes: Sodium, Potassium, Chloride, and Bicarbonate

Electrolytes must be in balance in order for cells and organs to function properly. When electrolytes are out of balance, the result is disorders and disease.

This chapter describes the physiology of sodium (Na), potassium (K), chloride (Cl) and bicarbonate (HCO_3^-), as well as how they are regulated, distributed, and measured in the body, including reference ranges. The chapter also discusses the conditions that result when these electrolytes are out of balance, and their causes and treatment.

OSMOLALITY AND VOLUME REGULATION

Physiology

Osmolality in plasma is an index of the number of solutes (soluble particles) dissolved in a kilogram of plasma water, with a normal osmolality between 280 and 300 mosm/kg. The relative osmolalities of blood plasma, cellular water, and interstitial fluid determine the water movements between each compartment and their volumes. By both active and passive transport, fluid moves between intra- and extracellular compartments through the cell walls, which function as semipermeable membranes. Water will pass through these cell walls from a compartment of lower osmolality into one of higher osmolality. This may be most serious clinically when a low plasma osmolality causes entry of excess fluid into the brain, causing cerebral edema and increasing intracranial pressure.

Serum osmolality is also one of several means to crudely assess overall body hydration, also called total body water (TBW). TBW is defined as the fluid that occupies the intracellular (~65% of TBW) and extracellular spaces (~10% intravascular; ~25% interstitial) of the body. (Note: Intravascular fluid has a much higher protein content than the interstitial fluid, with this protein exerting a water-retaining effect called the colloid osmotic pressure.) Regulating TBW involves complex balances between different fluid compartments in the body. In general, while TBW fluctuates hourly due to loss via lungs, skin, and kidneys and gain by food and fluid intake, plasma osmolality is more closely regulated and remains relatively constant *(1,2)*.

Changes in plasma osmolality are affected by numerous stimuli, including osmoreceptors in the hypothalamus that respond and regulate plasma osmolality. The regulation of osmolality also affects the sodium concentration in plasma, largely because sodium and its associated anions account for ~90% of the osmotic activity in plasma. The sodium concentration in blood is also affected by regulation of blood volume. As we will see later, although osmolality and volume are regulated by separate mechanisms [except for antidiuretic hormone (ADH), also called vasopressin], they are related because osmolality (sodium) is regulated by changes in water balance, whereas volume is regulated by changes in sodium balance *(3,4)*. In a person with normal water and osmolality homeostasis, if a large volume of hypotonic fluid, such as water, is consumed, the kidneys respond by rapidly excreting dilute urine, even before the intra- and extracellular fluids equilibrate and even if the body is dehydrated! This mechanism prevents fluid overload. In a healthy person, plasma osmolality does not respond to a decrase in TBW until TBW has decreased by 5% *(2)*.

The osmolality of plasma results from ions (Na, K, Cl, albumin, etc.) and neutral solutes such as glucose and urea. The contribution of the ions can be fortuitously estimated as $2 \times [Na]$ (in mmol/L), for three reasons:

- NaCl in plasma is 75% "osmotically active." Therefore, 1 mmol of NaCl behaves as if it dissociates into 1.75 mmol of osmotic particles (0.75 Na^+ + 0.75 Cl^- + 0.25 NaCl) *(3)*.
- Plasma is 93% water and 7% proteins and lipids (with ions mostly confined to the plasma H_2O space).
- Other ions such as K^+, Ca^{2+}, and Mg^{2+} account for ~8% of the osmotic activity relative to sodium ion.

Therefore, sodium contribution to osmolality is

$$(1.75/0.93) \times 1.08 \times [Na] = 2 \times [Na]$$

Calculated osmolality (mOsm/kg) =

$$(2 \times [Na]) + [BUN (mg/dL)/2.8] + [glucose (mg/dL)/18]$$

where BUN is blood urea nitrogen.

Note: Because concentrations of sodium, urea, and glucose are in molarity, calculations of "osmolality" are actually calculations of osmolarity.

Regulation of Osmolality

To maintain a normal plasma osmolality [~275–300 mmol/kg (~275–300 mOsm/kg) of plasma H_2O)], both thirst and ADH are activated. Osmoreceptors in the hypothalamus respond quickly to small changes in osmolality: A 1–2% increase in osmolality causes a fourfold increase in the circulating concentration of ADH, and a 1–2% decrease in osmolality shuts off ADH production entirely. ADH acts by increasing the reabsorption of water in the cortical and medullary collecting tubules. ADH has a half-life in the circulation of only 15–20 minutes.

Renal water regulation by ADH and thirst each play important roles in regulating plasma osmolality. Renal water excretion is more important in controlling water excess, whereas thirst is more important in preventing water deficit or dehydration. Consider what happens in the following conditions:

Water load

During excess intake of water (either done experimentally or in a condition such as polydipsia) before plasma osmolality declines, both ADH and thirst are suppressed. In the absence of ADH, a very large volume of dilute urine can be excreted (10 L or more per day), well above any normal intake of water. Therefore, hypo-osmolality and hyponatremia occur almost exclusively in patients with excess ADH and/or impaired renal excretion of water *(3)*.

Water deficit

As a deficit of water begins to increase plasma osmolality, both ADH secretion and thirst are activated. Even though ADH minimizes renal water loss, thirst is the major defense against hyperosmolality and hypernatremia because it stimulates a person to seek an external source of water. As an example of the effectiveness of thirst in preventing dehydration, a patient with diabetes insipidus (no ADH) may excrete 10 L of urine per day. However, because thirst persists, water intake matches output and plasma sodium remains normal *(3)*.

This is why hypernatremia rarely occurs in a person with a normal thirst mechanism and access to water. However, it becomes a concern in infants, unconscious patients, or anyone who is unable to either drink or ask for water. In people over the age of 60, osmotic stimulation of thirst diminishes. Particularly in the older patient with illness and diminished mental status, dehydration becomes increasingly likely.

Regulation of Blood Volume

Adequate blood volume is essential to maintain blood pressure and ensure good perfusion to all tissues and organs. Regulation of both sodium and water are interrelated in controlling blood volume. The renin-angiotensin-aldosterone system responds primarily to a decreased blood volume.

- Renin is secreted near the renal glomeruli in response to decreased renal blood flow (decreased blood volume and/or blood pressure). Renin converts angiotensinogen to angiotensin I, which then becomes angiotensin II.
- Angiotensin II has several actions. It stimulates thirst, causes vasoconstriction to quickly increase blood pressure, and causes secretion of aldosterone.
- Aldosterone acts to increase blood volume by increasing renal retention of sodium and the water that accompanies the sodium.

The effects of blood volume and osmolality on sodium and water metabolism are shown in Figure 3-1.

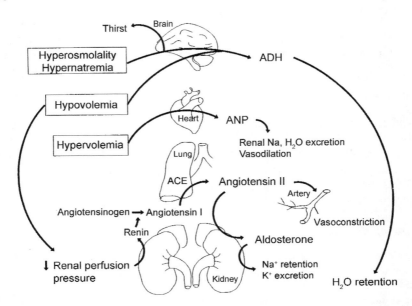

FIGURE 3-1. Effects of changes in blood volume and osmolality. Hypovolemia stimulates both the renin-angiotensin-aldosterone system and secretion of ADH, with ADH being the one common response to both hypovolemia and hyperosmolality. Volume receptors located in the afferent renal arterioles, the heart, the arteries, and the distal tubule, by sensing decreased blood pressure, stimulate the secretion of renin. Note that angiotensin-converting enzyme (ACE) is located in tissues other than the lung, such as the glomeruli and brain *(1). Adapted with permission from Kaplan LA, Pesce AJ. Clinical chemistry: theory, analysis and correlation, 1st ed. St. Louis: CV Mosby, 1984:371.*

Changes in blood volume (actually, pressure) are initially detected by a series of stretch receptors in areas such as the cardiopulmonary circulation, the carotid sinus, the aortic arch, and the glomerular arterioles. These receptors then activate a series of responses (effectors) that restore volume by appropriately varying vascular resistance, cardiac output, and renal sodium and water retention *(3,4)*. There are many effectors of blood volume:

- Secretion of epinephrine and norepinephrine are inversely related to volume changes.
- Angiotensin II is produced in response to decreased blood volume and leads to vasoconstriction, increased renal reabsorption of sodium, and release of aldosterone.
- Aldosterone promotes distal tubular reabsorption of Na^+ and Cl^- in exchange for K^+ and H^+.
- Natriuretic peptides (ANP and the now-famous BNP) are released from the myocardium in response to volume expansion and promote sodium excretion in the kidney.

Electrolytes: Sodium, Potassium, Chloride, and Bicarbonate

- Both thirst and ADH are stimulated by low blood volume, independently of osmolality.
- Glomerular filtration rate increases with volume expansion and decreases with volume depletion.
- All other things being equal, an increased plasma sodium will increase urinary sodium excretion, and vice versa.

The renal retention of sodium has a profound effect on blood volume, because whenever a sodium ion is reabsorbed, a water molecule follows along. While large amounts of sodium are filtered in the 150 L of glomerular filtrate produced daily, the renal tubules reabsorb 98–99% of this sodium along with most of the water. A 1–2% reduction in tubular reabsorption of sodium can increase water loss by several liters per day.

Urine osmolality values may vary widely depending on water intake and the circumstances of collection. However, it is generally decreased in diabetes insipidus (inadequate ADH) and polydipsia (excess H_2O intake due to chronic thirst) and increased in conditions in which ADH secretion is increased by hypovolemia or hyperosmolality or conditions such as inappropriate ADH.

Reference Ranges for Osmolality

Reference ranges for osmolality are shown in Table 3-1.

SODIUM

Physiology and Regulation

As an electrolyte, sodium helps transmit nerve impulses and activate muscle movements. Sodium is the most abundant cation in the ECF, representing

TABLE 3-1. Reference Ranges for Electrolytes and Osmolality[a]

	SI units	Conventional units
Sodium	135–145 mmol/L	135–145 mEql/L
Potassium	3.5–5.0 mmol/L	3.5–5.0 mEql/L
Chloride	98–107 mmol/L	98–107 mEq/L
Total CO_2	21–28 mmol/L	21–28 mEql/L
Anion gap		
Na, Cl, HCO_3^-	8–16 mmol/L	8–16 mEql/L
Na, K, Cl, HCO_3^-	11–20 mmol/L	11–20 mEq/L
Osmolality, plasma		
Child, adult	275–295 mmol/kg	275–295 mOsm/kg
Adult >60 y	280–300 mmol/kg	280–300 mOsm/kg
Osmolality, urine (24-h collection)	300–900 mmol/kg	300–900 mOsm/kg

[a] All values are plasma values unless noted otherwise.

90% of all extracellular cations, and largely determines the osmolality of the ECF. To maintain the much higher sodium concentration in the ECF relative to the intracellular concentration, an active transport system exists in cells. Since potassium is the major intracellular cation, it is continually exchanging with sodium. To maintain this large gradient between ECF and ICF, a transport mechanism involving a Na-K ATPase pump exchanges three sodium ions moving out of cells for two potassium ions moving into cells *(4)*. In simple terms, if the sodium in plasma gets too low (hyponatremia), water moves into cells, causing them to swell. This is especially dangerous for brain cells because, being confined within the skull, their expansion increases intracranial pressure.

The plasma sodium concentration depends greatly on the intake and excretion of water and, to a somewhat lesser degree, the renal regulation of sodium. Three processes are of primary importance:

- The intake of water in response to thirst, as stimulated or suppressed by plasma osmolality
- The excretion of water, largely affected by ADH release in response to changes in either blood volume or osmolality
- The blood volume status, which affects sodium excretion through aldosterone, angiotensin II, and ANP

The kidneys can conserve or excrete large amounts of sodium, depending on the osmolality (mostly sodium) of the ECF and the blood volume. Normally, the kidney reabsorbs 98–99% of filtered sodium: 60–75% is reabsorbed in the proximal tubule, with the remainder reabsorbed in the loop and distal tubule and exchanged for potassium in the connecting segment and cortical collecting tubule (under the control of aldosterone).

Evaluation of Hyponatremia

Hyponatremia is defined as a plasma sodium concentration <135 mmol/L, with its duration also an important consideration in treatment *(5)*. At values below 120 mmol/L, general weakness and mental confusion are typical. Paralysis and severe mental impairment may occur at plasma sodium <110 mmol/L *(6)*. Cerebral edema may lead to respiratory insufficiency and hypoxia, which can cause death *(7)*.

Most patients with hyponatremia have an excess of ADH, either from an inappropriate secretion of ADH or a decreased blood volume *(8)*. History is an important part of evaluating patients with hyponatremia, with heart failure, cirrhosis, GI losses, burns, endocrine disorders, excessive sweating, and drugs being possible causes. Chronic hyponatremia is a common clinical problem in the elderly, especially women *(9)*.

Here are the clinical and laboratory parameters that may be useful in the initial evaluation of hyponatremia:

Electrolytes: Sodium, Potassium, Chloride, and Bicarbonate

History

- Duration of the hyponatremia
- Edema (heart failure, cirrhosis)
- GI losses
- Skin losses (burns, sweating)
- Renal losses (low aldosterone, diuretics, salt-losing syndromes)
- Drugs (carbamazepine, SSRIs)
- Cancer (oat-cell carcinoma of lung typical cause of SIADH)

Clinical

- Check blood pressure and examine skin:
  Volume depletion: Decreased BP; skin has decreased turgor, and is cool and pale
  Volume overload: Elevated BP, pitting edema in lower extremities

Laboratory

- Plasma Na, K, Cl, and HCO_3^-
- Anion gap (calculated from the above parameters)
- Plasma urea, glucose, and uric acid
- Urine osmolality and Na

The diagnostic approach to hyponatremia is shown in Table 3-2. Decreased plasma Na should be confirmed by a decreased plasma osmolality. Plasma urea and uric acid are often low with inappropriate secretion of

TABLE 3-2. Differential Diagnosis of Hyponatremia

Measure plasma Na, K, Cl, HCO_3^-, and urea; and osmolality and uric acid if needed. Plasma Na and osmolality decreased.
Measure urine Na:
  Urine Na <15 mmol/L (<15 mEq/L); plasma AG, urea, and uric acid normal to increased.
    Hypovolemia with hypotonic fluid replacement (diarrhea, vomiting, sweating, renal losses, etc.)
    Polydipsia (chronic thirst)
    Hypervolemia with arterial hypovolemia (congestive heart failure, cirrhosis)

  Urine Na >20 mmol/L (>20 mEq/L); plasma AG, urea, and uric acid normal to decreased.
    Inappropriate excess secretion of ADH (carcinoma of lung, adrenal insufficiency, etc.)
    Renal salt loss (thiazides, aldosterone deficiency)
    Reset osmostat
    Renal failure with water overload

ADH *(8)*. The urine Na and urine osmolality help differentiate among the causes of hyponatremia.

The pathogenesis of hyponatremia is often related to volume status and excess ADH that is secreted in response to a volume or osmotic stimulus, as shown in Table 3-3 and Figure 3-2. Volume status may be assessed by skin turgor, jugular venous pressure, and urine Na concentration, with a low urine Na indicating hypovolemia *(1,3,4)*.

Hypovolemic hyponatremia occurs when both water and sodium levels in blood are decreased as a result of sodium loss in excess of water loss. There are several common causes of this condition:

- Use of thiazide diuretics (but not loop diuretics) induces sodium and potassium loss without interfering with ADH-mediated water retention.
- As hypotonic fluid is lost by prolonged vomiting, diarrhea, or sweating, it is replaced by relatively more hypotonic fluid if water is ingested in response to thirst and ADH is stimulated by hypovolemia (see Figure 3-2).
- In potassium depletion, cellular loss of potassium promotes sodium movement into the cell with an associated decrease in plasma Na and volume.
- Aldosterone deficiency increases urinary sodium loss (may also be normovolemic).
- Salt-wasting nephropathy may infrequently develop in renal tubular and interstitial diseases, such as medullary cystic and polycystic kidney diseases, usually as renal insufficiency becomes severe [serum creatinine >610 μmol/L (>8 mg/dL)].

TABLE 3-3. Hyponatremia Related to Volume Status

Hypovolemic	**Na loss in excess of H_2O**
	Thiazide diuretics
	Loss of hypertonic fluid: GI, burns, sweat
	Potassium depletion
	Aldosterone deficiency
	Salt-losing nephropathies
Euvolemic	**Problem with water balance**
	Excess or inappropriate ADH secretion
	Artifactual—severe hyperlipidemia
	Hyperosmolar from substance other than sodium
	Polydipsia
	Adrenal insufficiency
	Altered regulatory set point for osmolality
	Drugs
Hypervolemic	**Movement of fluid from intravascular to interstitial space**
	Congestive heart failure, hepatic cirrhosis
	Advanced renal failure (decreased glomerular filtration rate) with excess water intake
	Nephrotic syndrome—decreased colloid osmotic pressure

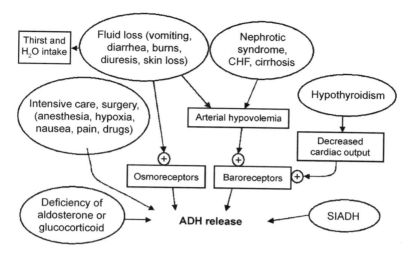

FIGURE 3-2. Conditions that lead to ADH release and possible hyponatremia. CHF, congestive heart failure; SIADH, syndrome of inappropriate secretion of antidiuretic hormone.

Normovolemic hyponatremia typically indicates a problem with water balance and may be related to one of the following circumstances:

- Polydipsia (chronic thirst with excess intake of water) eventually leads to hyponatremia that is usually mild but occasionally severe.
- Excess ADH may be secreted in response to drugs, surgery, tumors, central nervous system disorders, endocrine disorders, or pulmonary conditions *(4,8)*. The excess ADH causes mild hypervolemia, which then leads to excretion of sodium and water by release of ANP (see Figure 3-2).
- Artifactual hyponatremia can occur in cases of severe hyperlipidemia or hyperproteinemia. Methods that dilute plasma before sodium analysis by flame photometry or ISE give erroneously low Na results on these samples because they measure millimoles of Na per liter of plasma. "Direct" methods by ISE, which do not dilute plasma or whole blood, give accurate Na results because they detect the sodium concentration in the plasma water.
- Severe hyperglycemia generally causes a lower plasma sodium in order to maintain a normal plasma osmolality.
- In adrenal insufficiency, the decreased aldosterone (mineralocorticoid) and cortisol (glucocorticoid) promote ADH release *(3)*.
- In pregnancy, osmotic regulation in the hypothalamus may be offset such that plasma Na is regulated ~5 mmol/L (~5 mEq/L) lower than normal. This effect on the hypothalamus may be initiated by vasodilation, which mimics hypovolemia by lowering blood pressure.
- Drugs: desmopressin, psychoactive agents, anti-cancer agents *(12)*.

Hypervolemic hyponatremia is nearly always a problem of water overload as excess water accumulates from renal failure, heart failure, or liver failure. It is usually associated with one or more of the following conditions:

- Excess secretion of ADH. For example, congestive heart failure or hepatic cirrhosis increases venous back-pressure in the circulation, which promotes movement of fluid from the blood to the interstitium, causing edema and arterial hypovolemia. The arterial hypovolemia stimulates ADH secretion, which eventually leads to hypervolemia and hyponatremia, as indicated in Figure 3-2.
- Inability of the kidneys to excrete excess water along with excess fluid intake. This is more likely in elderly patients.

Treatment of Hyponatremia

The treatment of hyponatremia depends on the cause, the severity, and how rapidly the hyponatremia develops *(5,10,12)*. While fluid restriction is effective in some patients *(11)*, a study of postmenopausal women with chronic symptomatic hyponatremia showed significantly better outcomes with giving hypertonic saline vs. fluid restriction *(8,9)*. Because all patients without hypoxia recovered, hypoxia was believed to be a mechanism for brain damage.

Acute hyponatremia developing within 48 hours carries a greater risk of cerebral edema and should be treated aggressively *(5)*. It may require intravenous (iv) administration of hypertonic sodium solutions. However, correction should be no faster than about 10 mmol/L per day *(12)*. Note that these rates are not *goals* of therapy; they are upper limits not to be exceeded *(5)*. If hyponatremia is chronic, it should generally be corrected slower than 8 mmol/L/24h. If convulsions or coma is present, a rapid initial correction with hypertonic saline may be necessary to lower the hyponatremia by up to 4 mmol/L to reverse the symptoms, followed by a more gradual correction *(5)*.

Hypervolemic hyponatremia or asymptomatic hyponatremia is usually treated effectively with water restriction. Also, in advanced renal disease, the inability to excrete water promotes hypervolemia whenever fluid intake is excessive.

Evaluation of Hypernatremia

Hypernatremia (increased serum sodium concentration) usually results from excessive loss of water relative to sodium. Hypotonic fluid may be lost through several common routes: by the kidney, by profuse sweating, or by GI loss such as diarrhea. About 1 L of water per day is normally lost through the skin and by breathing (insensible losses). Any condition that increases water loss, such as fever, burns, or exposure to heat, will increase the likelihood of developing hypernatremia. However, hypernatremia rarely occurs in persons with a normal thirst mechanism and access to water. Severe symptoms usually require an acute increase in sodium concentration to 158 mmol/L or higher *(12)*. Cerebral dehydration can lead to demyelination, cerebral bleeding, coma, and death *(10)*. Because the brain can adapt to the slower (chronic) onset of hypernatremia by restoring normal cell volume, chronic hypernatremia is less likely to induce neurologic symptoms *(12)*. Hypernatremia should be corrected at a rate no faster than about 0.5 mmol/L/h.

The measurement of urine osmolality is necessary to evaluate the cause of hypernatremia. Interpretation of the urine osmolality in hypernatremia is shown in Table 3-4. In renal loss, the urine osmolality is low or normal. In nonrenal losses, the urine osmolality is increased.

Nonrenal causes of hypernatremia are very common, with examples as follows:

- Inadequate fluid intake to replace water that is lost continuously by breathing, sweating, evaporation through skin, or by GI loss of hypotonic fluid.
- Altered mental status in elderly patients and in infants, both of whom may have an intact thirst mechanism, but who are unable to ask for or obtain water.
- Neonates are also susceptible to hypernatremia caused by iv administration of excess hypertonic saline.
- Diminished or absent thirst reflex (hypodipsia). Also, thirst normally diminishes in the elderly.
- Ingestion of too much salt is an obvious but uncommon primary cause of chronic hypernatremia. In the nonrenal examples above, ADH secretion is increased to minimize renal loss of water, resulting in urine osmolality >800 mmol/kg (>800 mOsm/kg) (Table 3-4).

TABLE 3-4. Evaluation of Hypernatremia[a]

If plasma Na is >150 mmol/L (150 mEq/L), measure urine osmolality.
>800 mmol/kg (>800 mOsm/kg) (inadequate fluid intake)
    Insensible losses of water
    GI loss of hypotonic fluid
    Diminished or lost thirst or inability to obtain water
    Excess iv or oral intake of Na
300–800 mmol/kg (300–800 mOsm/kg)
    Impaired ADH release (hyperaldosteronism, Cushing's syndrome)
    Partial central diabetes insipidus
    Diuretics
    Osmotic diuresis
<300 mmol/kg (<300 mOsm/kg)
    Diabetes insipidus (central or nephrogenic)

[a]Persons who cannot fully concentrate their urine, such as neonates, young children, the elderly, and some patients with renal insufficiency, may show a relatively lower urine osmolality.

Renal causes of hypernatremia are often related to insufficient ADH, which is a general mechanism for hypernatremia by increasing renal water loss. Hyperglycemia can elevate osmolality simply by the presence of glucose in the blood, and may also cause hypernatremia by glucose-mediated loss of hypotonic fluids during osmotic diuresis.

Chronic hypernatremia in an alert patient with a normal thirst is indicative of hypothalamic disease. This includes the following:

- Primary hyperaldosteronism, in which excess aldosterone retards ADH release and induces a mild hypervolemia
- Cushing's syndrome, with excess production of cortisol *(3,4)*
- ADH deficiency may be caused by impaired ADH secretion (diabetes insipidus) or by impaired renal response to ADH (nephrogenic diabetes insipidus). Because increased thirst typically compensates for renal water loss, hypernatremia does not usually occur in diabetes insipidus unless the thirst mechanism is also impaired. Diabetes insipidus is characterized by copious production of dilute urine (3–20 L/d). A partial defect in either ADH release or ADH effect at the receptor level may also occur. In such cases, the kidneys are unable to sufficiently concentrate the urine to correct the hypernatremia.
- Excess water loss may also occur in renal tubular disease, such as acute tubular necrosis, in which the tubules become unable to fully concentrate the urine, causing an osmotic diuresis.

Treatment of Hypernatremia

In patients who have developed hypernatremia acutely over a period of hours, rapid correction of the hypernatremia is beneficial, while minimizing the risk of cerebral edema *(12)*. Because too-rapid correction of serious chronic hypernatremia (>160 mmol/L) can induce cerebal edema and death, hypernatremia must be corrected gradually: Typically, acute hypernatremia can be corrected at about 1 mmol/L per hour, while chronic hypernatremia should be corrected at a maximal rate of 0.5 $mmol \cdot L^{-1} \cdot h^{-1}$ until the plasma Na reaches 145 mmol/L *(3,12)*.

Reference Ranges for Sodium

Reference ranges for sodium are shown in Table 3-1.

POTASSIUM

Physiology

Potassium is the major intracellular cation, with a 20-fold greater concentration in the cells than in the ECF (~4 mmol/L). Cellular membrane potentials and other functions require that the body maintain the appropriate ratio of potassium concentrations between the cell and ECF. Insulin, B-adrenergic catecholamines, thyroid hormone, and acid-base balance influence this cellular transport. Only 2% of the total potassium in the body circulates in the plasma.

Potassium has several vital functions, including *(3)*

- Regulation of neuromuscular excitability
- Contraction of the heart and cardiac rhythm
- Regulation of intracellular and extracellular volume and acid-base status

The potassium ion concentration in plasma has a major effect on the contraction of cardiac muscle. An increased plasma potassium slows the heart rate by decreasing the resting membrane potential of the cell relative to the threshold potential. A decrease in the plasma potassium concentration to <3.5 mmol/L increases the likelihood of arrhythmias. More severe hypokalemia causes general muscle weakness and more severe arrhythmias and atrial tachycardia. At potassium levels of 2.5 mmol/L, myopathy may progress to rhabdomyolysis with myoglobinuria. Also, both renal failure and respiratory function may be impaired *(12)*.

The potassium concentration also affects the H^+ concentration in the blood. When hypokalemia is present, potassium ions are lost from the cells; as a result, sodium and hydrogen ions move into the cell to replace potassium. The H^+ concentration is therefore decreased in the ECF, resulting in alkalemia *(4)*. The reverse is true for hyperkalemia.

Regulation

The kidneys are important in the regulation of potassium balance. Initially, the proximal tubules reabsorb nearly all the potassium. Then, under the influence of aldosterone, additional potassium is secreted into the urine in exchange for sodium in both the distal tubules and the collecting ducts. Thus, the distal nephron is the principal determinant of urinary potassium excretion.

An acute oral or iv intake of potassium is handled by potassium uptake from the ECF into the cells. Excess plasma K rapidly enters cells to normalize plasma K. As the cellular potassium then gradually returns to the plasma, it is removed by urinary excretion. Note that chronic loss of cellular potassium may result in cellular depletion without appreciable change in the plasma K concentration, because any excess potassium is normally excreted in the urine.

There are several important factors that influence the distribution of potassium between cells and ECF:

- Potassium loss frequently occurs whenever the Na-K ATPase pump is inhibited by conditions such as hypoxia, hypomagnesemia, or digoxin overdose.

- Insulin promotes acute entry of K ions into skeletal muscle and liver by increasing Na-K ATPase activity.
- Catecholamines are also factors; for example, β_2-stimulators, such as epinephrine, promote cellular entry of potassium, whereas β-blockers, such as propranolol, impair cellular entry of potassium.

As mentioned earlier, alkalemia is associated with a mild decrease in the plasma potassium concentration. As cells release H^+ to normalize alkalemia, extracellular K^+ and Na^+ enter the cell to maintain electroneutrality. Hypokalemia can also induce alkalosis by several mechanisms, such as reabsorption of HCO_3^- and increasing distal nephron secretion of acid *(4)*.

Causes of Hypokalemia

Hypokalemia is defined as a plasma potassium concentration <3.5 mmol/L (<3.5 mEq/L). Hypokalemia can occur with

- GI or urinary loss of potassium
- Increased cellular uptake of potassium
- Chronically reduced dietary intake (rare as the sole cause)

On average, serum potassium decreases by 0.3 mmol/L for each 100 mmol loss from body stores.

GI and renal losses

GI loss of potassium most often occurs through vomiting, diarrhea, laxative use, gastric suction, or discharge from an intestinal fistula. Renal loss of potassium can result from kidney disorders such as renal tubular acidosis, potassium-losing nephritis, and diabetic ketoacidosis. Excess aldosterone can lead to hypokalemia and metabolic alkalosis *(3,4)*. Hypomagnesemia can lead to hypokalemia by promoting both urinary and fecal loss of potassium. Magnesium deficiency diminishes the activity of Na-K ATPase and enhances the secretion of aldosterone. Effective treatment requires supplementation with both magnesium and potassium *(3)*. Reduced dietary intake of potassium is a rare cause of hypokalemia in healthy persons. However, decreased intake may intensify hypokalemia caused by other factors, such as use of diuretics.

Increased cellular uptake

Alkalemia, insulin, hypothermia, and a variety of drugs (epinephrine, bronchodilators, caffeine, etc.) can increase cellular uptake of potassium *(12)*. When blood is alkaline (alkalemia), intracellular H^+ moves into blood to normalize extracellular pH. As this happens, both potassium and sodium enter cells to maintain electroneutrality. Plasma K decreases by ~0.4 mmol/L

($\sim$0.4 mEq/L) per 0.1 unit rise in pH. Additional factors such as HCO_3^- administration, diuretics, and vomiting can intensify the hypokalemia *(3)*.

Insulin promotes the entry of potassium into skeletal muscle and liver cells. Insulin therapy may expose or intensify an underlying hypokalemic state *(3)*. Plasma K should be monitored carefully whenever insulin is administered to susceptible patients.

Clinical Effects of Hypokalemia

Hypokalemia is a serious concern in all patients but especially so in those with cardiovascular disorders. Hypokalemia is characterized by

- Muscle weakness
- Irritability
- Paralysis
- Serious cardiac abnormalities

Muscle weakness and paralysis are probably caused by alterations in polarization of the cell membrane. Muscle weakness and paralysis may also interfere with breathing.

A variety of cardiac arrhythmias can be induced when potassium falls below 2.5 mmol/L (2.5 mEq/L):

- Premature atrial and ventricular beats
- Sinus bradycardia
- Atrioventricular block
- Ventricular tachycardia and fibrillation

These may cause sudden cardiac arrest in some patients. Prior to surgery, hypokalemia [<3.5 mmol/L (<3.5 mEq/L)] has been associated with an increased incidence of arrhythmia and need for cardiopulmonary resuscitation *(13)*. Severe hypokalemia requires aggressive iv replacement with close laboratory monitoring. Potassium chloride or potassium phosphate is the usual salt administered. Potassium bicarbonate may be considered only when metabolic acidosis is present *(12)*.

Treatment of Hypokalemia

Treatment for hypokalemia includes oral or iv replacement of potassium. Especially in severe hypokalemia, other electrolytes may be lost, such as sodium, chloride, magnesium, bicarbonate, and acid. Replacement of these electrolytes should be guided by monitoring their blood levels. In some cases, chronic mild hypokalemia may be corrected simply by an intake of food with a high potassium content, such as bananas and orange juice in the diet.

Causes of Hyperkalemia

The most common causes of hyperkalemia are

- Excessive oral or iv intake of potassium
- Processes that enhance potassium release from cells to the ECF
- Decreased renal excretion

Patients with hyperkalemia often have an underlying disorder such as renal insufficiency, diabetes mellitus, or metabolic acidosis that, in combination with another process, causes hyperkalemia *(4,6,14)*. For example, during administration of KCl, a person with renal insufficiency is far more likely to develop hyperkalemia than is a person with normal renal function.

Increased load

Increased oral or iv intake of potassium commonly causes hyperkalemia if urinary K excretion is impaired. In healthy persons, an acute oral load of potassium will increase plasma K only transiently, because most of the absorbed potassium will rapidly move intracellularly. Normal cellular processes gradually release this excess potassium into the plasma, where it is removed by renal excretion. If potassium is infused intravenously at rates in excess of 20 mmol of K/h, hyperkalemia is especially likely if renal function is impaired.

Cellular losses or shifts

Potassium may be released into the ECF when tissue breakdown or catabolism is enhanced, resulting in hyperkalemia, especially if renal insufficiency is present *(12)*. Increased cellular breakdown is associated with these events:

- Trauma
- Administration of cytotoxic agents
- Severe tissue hypoxia
- Massive hemolysis
- Insulin deficiency
- Metabolic acidosis
- Blood transfusions
- Exercise
- Hyperosmolality

Many of the processes listed above produce extreme biochemical or physical stresses to cells, which releases large amounts of potassium. These stresses include severe trauma, cytotoxic drugs, severe tissue hypoxia, and some hemolytic processes. Patients on cardiac bypass may develop mild elevations in plasma potassium during warming after surgery. Hypothermia increases movement of potassium into cells, whereas warming causes cellular release of potassium.

In diabetes mellitus, insulin deficiency promotes cellular loss of potassium. Hyperglycemia also contributes by producing a hyperosmolar plasma that pulls water and some associated potassium from cells.

In metabolic acidosis, H^+ moves into cells, exchanging with K^+, leaving the cell to maintain electroneutrality. Plasma K increases by 0.2–1.7 mmol/L (0.2–1.7 mEq/L) for each 0.1 unit reduction in pH *(3)*. Because cellular potassium often becomes depleted in cases of acidosis with hyperkalemia (including diabetic ketoacidosis), treatment with agents such as insulin and bicarbonate can cause a rapid intracellular movement of potassium, producing severe hypokalemia.

In blood stored for transfusions, potassium is gradually released from erythrocytes during storage, which can markedly elevate plasma K concentrations after storage for several weeks. While this has little effect in adults with adequate renal function, infants are susceptible to hyperkalemia from this cause.

Exercise is uniformly associated with a rise in plasma K concentration, because potassium is released from cells during exercise. The rise is directly related to the intensity of exercise *(4)*: mild to moderate exercise may increase plasma K by 0.3–1.2 mmol/L (0.3–1.2 mEq/L), while exhaustive exercise may increase plasma potassium by >2.0 mmol/L (>2.0 mEq/L). These changes usually reverse after sufficient rest. Note that forearm exercise during venipuncture can cause erroneously high plasma K concentrations.

Hyperosmolality causes diffusion of water out of cells that carries intracellular potassium ions with water into the blood.

Decreased renal excretion

When glomerular filtration or tubular function is impaired, hyperkalemia is much more likely. A variety of conditions impair renal excretion of potassium by diminishing aldosterone production, such as hypoaldosteronism or Addison's disease (decreased cortisol) *(6)*, although one-third of patients with Addison's disease do not have hyperkalemia *(5)*.

Several drugs commonly cause hyperkalemia by inhibiting the production or effect of aldosterone, especially in patients with renal insufficiency. These drugs include

- Captopril and other inhibitors of angiotensin-converting enzyme
- Nonsteroidal anti-inflammatory agents (inhibit aldosterone)
- Spironolactone (potassium-sparing diuretic that blocks distal tubular secretion of potassium)
- Digoxin (inhibits Na-K pump)
- Cyclosporine (inhibits renal response to aldosterone)
- Heparin therapy (inhibits aldosterone secretion)

Clinical Effects of Hyperkalemia

The clinical effects of hyperkalemia depend largely on the rate of increase in serum potassium level. Hyperkalemia can cause muscle weakness by decreasing the ratio of intra- to extracellular potassium, which alters neuromuscular conduction. Muscle weakness does not usually develop until plasma potassium reaches 7 mmol/L (7 mEq/L), although the correlation between symptoms and potassium levels varies among patients *(3,12)*.

Hyperkalemia disturbs cardiac conduction, which can lead to cardiac arrhythmias and possible cardiac arrest. Plasma potassium concentrations of 6–7 mmol/L (6–7 mEq/L) may alter the electrocardiogram *(3)*, and concentrations >10 mmol/L (>10 mEq/L) may cause cardiac arrest *(12)*.

Treatment of Hyperkalemia

By reducing the threshold potential of myocardial cells, calcium offsets the effect of potassium, which lowers the resting potential of myocardial cells. Therefore, calcium given as calcium gluconate provides immediate but short-lived protection to the myocardium against the effects of hyperkalemia *(12)*.

Insulin may be given to activate the Na-K-ATPase to promote potassium movement back into cells. Glucose is also administered to prevent hypoglycemia. Infusion of sodium bicarbonate will lower serum potassium levels and may be effective in patients with metabolic acidosis. Patients treated with these agents must be carefully monitored to prevent hypokalemia as potassium moves back into cells. A cation-exchange resin, such as sodium polystyrene sulfonate, lowers plasma K by binding potassium ions in the gut. If the preceding measures do not prove successful or if renal failure occurs, peritoneal dialysis or, if more rapid removal is necessary, hemodialysis, may be used *(3,12)*.

Proper Collection and Handling of Samples

There are many causes of artifactual hyperkalemia:

- First, the coagulation process releases potassium from platelets, so that serum K may be 0.1–0.5 mmol/L (0.1–0.5 mEq/L) higher than plasma K concentrations *(6)*. If the patient's platelet count is elevated (thrombocytosis), serum potassium may be further elevated. This situation may be avoided by using a heparinized tube to prevent clotting of the specimen.
- Second, if a tourniquet is left on the arm too long during blood collection, or if the patient excessively clenches the fist or otherwise exercises the forearm before venipuncture, cells may release potassium into the plasma. This can be prevented by using proper care in the drawing of blood.
- Third, because storing blood on ice promotes the release of potassium from cells *(15)*, whole blood samples for potassium determinations should never be iced, but instead stored at room temperature and analyzed promptly. Otherwise the blood should be centrifuged to remove the cells.
- Fourth, if hemolysis occurs after the blood is drawn, potassium will be falsely elevated, depending on the degree of hemolysis.

Reference Ranges for Potassium

Reference ranges for potassium are shown in Table 3-1.

CHLORIDE

Physiology and Regulation

As the major extracellular anion, Cl functions with Na, K, and other cations in conduction and transport mechanisms between cells and across cell membranes. Chloride has several important functions that are sometimes passive responses to other ion movements:

- It maintains osmotic and fluid balance along with Na.
- It is transported along with cations and exchanged with bicarbonate to maintain ionic neutrality.
- In acid-base balance, it has an inverse relationship with bicarbonate.

Regulation of Cl appears to be in direct response to regulation of other ions. Aldosterone indirectly increases Cl retention in the kidney by enhancing Na reabsorption in the renal tubules, with either K exchanging or Cl following the Na ion to balance the ionic charge movements *(16)*. In the ascending limb of the loop, a Cl pump actively reabsorbs Cl *(6)*. In the GI tract, Cl may passively follow as Na is absorbed, or Cl may exchange with HCO_3^- as HCO_3^- is secreted into the intestine. ADH deficiency, by promoting renal loss of H_2O, leads to increases of Na and Cl in blood. Elevated PTH levels are associated with increased serum Cl and may play an indirect role in Cl regulation *(17)*.

Chloride is normally lost through sweat, urine, and gastric secretions. Cl ion movements are related to acid-base balance, with Cl and HCO_3^- having an inverse relationship. In chloride deficiency, whenever sodium is reabsorbed, more HCO_3^- is preferentially reabsorbed instead of chloride. With primary aldosterone excess, increased tubular reabsorption of sodium is accompanied by both potassium and H loss and gain of HCO_3^-. According to the strong ion difference theory (see Chapter 1), the movement of Cl ion may directly produce the strong acid in gastric fluid. If Cl ion is transported across a membrane without an accompanying cation, one means to maintain ionic neutrality is for H_2O to dissociate into a H ion (which remains with chloride) and a OH ion (which is transported in the opposite direction).

Sweat Cl measurements have long been used to confirm the diagnosis of cystic fibrosis in children. Use of this test is increasing as screening requirements for cystic fibrosis in children have increased *(6)*.

Causes of Hypochloremia and Hyperchloremia

Hypochloremia occurs when Cl levels in blood drop below 98 mmol/L. Hypochloremia can be due to the following causes:

- Excessive GI loss of chloride due to loss of HCl by vomiting or gastric suctioning
- Renal loss due to prolonged use of diuretics or to tubular dysfunction that causes excess loss of Cl in urine. Hypochloremic metabolic alkalosis can result from these excess losses of Cl

- Aldosterone excess
- In chronic pulmonary disease with respiratory acidosis, as metabolic compensation increases HCO_3^- reabsorption, more Cl is lost *(4,16)*.

A low Cl in blood can cause metabolic alkalosis by enhancing renal HCO_3^- reabsorption.

Chloride measurements are useful when interpreting difficult acid-base disorders, because hyperchloremia often indicates an acidotic process (primary or compensatory), whereas hypochloremia indicates a primary or compensatory alkalotic process.

Hyperchloremia occurs when Cl levels in blood exceed 107 mmol/L. Hyperchloremia may occur with excessive loss of bicarbonate due to

- GI losses (such as diarrhea)
- Renal tubular acidosis
- Aldosterone deficiency
- ADH deficiency
- Compensated respiratory alkalosis
- Drugs, such as cortisone and acetazolamide

Hyperchloremia frequently accompanies hypernatremia. Mild elevations may also be seen in primary hyperparathyroidism *(17)*.

Reference Ranges for Chloride

Reference ranges for chloride are shown in Table 3-1.

BICARBONATE

Physiology and Regulation

HCO_3^- is the second most abundant anion in the extracellular fluid. Total CO_2 concentration normally ranges from 22 to 27 mmol/L and is composed of HCO_3^-, carbonic acid (H_2CO_3), and dissolved CO_2, with HCO_3^- accounting for >90% of the total CO_2 at physiological pH.

Depending on your belief, HCO_3^- is either the major determinant of our acid-base status or, according to Stewart's strong ion difference (SID) concepts, HCO_3^- is merely a variable dependent on pCO_2, SID, and weak acid concentrations *(18)*. Regardless, HCO_3^- is the major component of the buffering system in the blood. Carbonic anhydrase in erythrocytes converts CO_2 to HCO_3^-. To buffer against acidosis, HCO_3^- combines with excess H^+ to produce CO_2 and H_2O. The loss of excess CO_2 (an acidic gas) by the lungs gives the HCO_3^--CO_2 system a great capacity to buffer acid production. In the kidneys, most (85%) of the HCO_3^- is reabsorbed by the proximal tubules, with 15% being reabsorbed by the distal tubules.

Causes of Decreased and Increased Bicarbonate

HCO_3^- decreases in metabolic acidosis when HCO_3^- combines with H^+ to produce CO_2, which is exhaled by the lungs. The typical response to metabolic acidosis is compensation by hyperventilation, which lowers pCO_2. Some typical causes of metabolic acidosis are

- Hypoxia
- Ketoacidosis
- Diarrhea

Elevated total CO_2 concentrations occur in metabolic alkalosis as bicarbonate is retained, often with an increase in pCO_2 due to compensation by hypoventilation. Typical causes of metabolic alkalosis include (19)

- Severe vomiting
- Hypokalemia
- Excessive alkali intake

Reference Ranges for Bicarbonate

Reference ranges for bicarbonate are shown in Table 3-1.

REFERENCES

1. Nebelkopf Elgart H, Johnson KL, Munro N. Assessment of fluids and electrolytes. Amer Assoc Crit-Care Nurses: advanced practice in critical care 2004;15:607–21.
2. Armstrong LE. Assessing hydration status: the elusive gold standard. J Amer College Nutrition 2007;26:575S–84S.
3. Rose BD. Clinical physiology of acid-base and electrolyte disorders, 3rd ed. New York: McGraw-Hill, 1989.
4. Narins RG, ed. Maxwell and Kleeman's clinical disorders of fluid and electrolyte metabolism, 5th ed. New York: McGraw-Hill, 1994.
5. Hoorn EJ, Halperin ML, Zietse R. Diagnostic approach to the patient with hyponatremia: traditional versus physiology-based options. Q J Med 2005;98:529–40.
6. Scott MG, LeGrys VA, Klutts JS. Electrolytes and blood gases. In: Burtis CA, Ashwood ER, Bruns DE, ed. Tietz textbook of clinical chemistry and molecular diagnostics, 4th ed. St Louis: Elsevier Saunders, 2006:983–1018.
7. Knochel JP. Hypoxia is the cause of brain damage in hyponatremia. JAMA 1999;281:2342–3.
8. Decaux G, Musch W. Clinical laboratory evaluation of the syndrome of inappropriate secretion of antidiuretic hormone. Clin J Am Soc Nephrol 2008;3:1175–84.
9. Ayus JC, Arieff AI. Chronic hyponatremic encephalopathy in postmenopausal women: association of therapies with morbidity and mortality. JAMA 1999;281:2299–304.
10. Sedlacek M, Schoolwerth AC, Remillard BD. Electrolyte disturbances in the intensive care unit. Seminars in Dialysis 2006;19:496–501.

11. Lauriat S, Berl T. The hyponatremic patient: practical focus on therapy. J Am Soc Nephrol 1997;8:1599–607.
12. Weiss-Guillet E-M, Takala J, Jakob SM. Diagnosis and management of electrolyte emergencies. Best Practice & Research Clin Endocrinol Metab 2003;17:623–51.
13. Wahr JA, Parks R, Biosvert D, et al. Preoperative serum potassium levels and perioperative outcomes in cardiac surgery patients. JAMA 1999;281:2203–10.
14. Rimmer JM, Horn JF, Gennari FJ. Hyperkalemia as a complication of drug therapy. Arch Intern Med 1987;147:867–9.
15. Fleisher M, Gladstone M, Crystal D, et al. Two whole-blood multi-analyte analyzers evaluated. Clin Chem 1989;35:1532–5.
16. Powers F. The role of chloride in acid-base balance. J Intravenous Nursing 1999; 22:286.
17. Lafferty FW. Primary hyperparathyroidism: changing clinical spectrum, prevalence of hypertension, and discriminant analysis of laboratory tests. Ann Intern Med 1981;141:1761–6.
18. Story DA. Bench-to-bedside review: a brief history of clinical acid-base. Crit Care 2004;8:253–8.
19. Khanna A, Kurtzman NA. Metabolic alkalosis. J Nephrol 2006; suppl 9:S86–S96.

Appendix

Self-Assessment and Mastery

CHAPTER 1
Acid-Base Exercises

In the following table, use the pH, pCO_2, and HCO_3^- values along with the duration of observation to assess the acid-base status of each situation. Determine the primary disorder, then evaluate if the expected compensation has occurred for the duration indicated. Use Figure 1-3 for guidance as needed.

pH	pCO_2 (mmHg)	HCO_3^- (mmol/L)	Duration of medical observation	Acid-base condition
7.40	40	24	—	Very normal acid-base results
7.34	34	18	12 h	?
7.55	49	40	6 h	?
7.55	49	40	24 h	?
7.44	27	18	2 d	?
7.23	70	29	6 h	?
7.23	70	29	2 d	?
7.28	55	25	2 h	?
7.41	70	43	2 h	?
7.52	31	29	6 h	?
7.08	54	18	12 h	?
7.56	20	20	6 h	?

Cases:

Assess the following patients for their acid-base, ventilatory, and oxygenation status. Remember to assess the clinical picture for clues and determine the anion gap and delta ratio where appropriate.

Case 1

A 54-y-o woman with a history of hypertension and aortic valve stenosis developed severe edema and shortness of breath. She was admitted for aortic valve replacement. After this operation she remained on a ventilator for several days, but is now breathing 50% oxygen on her own. Her ventilatory mechanics were marginal to poor. Her arterial blood gas results were as follows:

pH   7.28 ACIDIC
pCO_2 53 mmHg ACID
pO_2 116 mmHg
HCO_3^- 25 mmol/L (25 mEq/L) okay
FI-O_2 (%)   50%

RESP. ACIDOSIS

Case 2

A female infant was operated on at 7 d of age to correct pulmonary atresia and a patent ductus arteriosus. A central shunt was placed from the right ventricle to the pulmonary artery. Several days after the operation, she developed congestive heart failure and pulmonary edema due to excess shunt flow. Bicarbonate was administered to correct lactate acidosis due to inadequate peripheral perfusion. At 11 d, the central shunt was revised with a smaller diameter shunt. At 10 and 15 d, she was breathing 40% oxygen when the following arterial blood gas results were obtained:

	Day 10	Day 15
pH	7.48	7.44
pCO_2	46 mmHg	44 mmHg
pO_2	40 mmHg	60 mmHg
HCO_3^-	34 mmol/L (33 mEq/L)	30 mmol/L (30 mEq/L)
Hb	102 g/L (10.2 g/dL)	108 g/L (10.8 g/dL)
Lactate	5.8 mmol/L (52 mg/dL)	1.5 mmol/L (13.5 mg/dL)

Case 3

A young boy with a history of asthma is admitted to the ED with a 2-d history of difficulty breathing and a breathing rate of 28/min (normal about 13/min). He is alert but is wheezing and has great difficulty speaking. Here are his lab results:

pH   7.40
pCO_2 28 mmHg
pO_2 45 mmHg
sO_2 70%
Na: 134 mmol/L
K   4.9 mmol/L
Cl   96 mmol/L
HCO_3^- 16 mmol/L (16 mEq/L)

Case 4

A 32-y-o man with diabetes and renal failure was admitted for renal transplantation. Intraoperative blood gas results were as follows, while he was breathing 60% oxygen:

pH	7.20
pCO_2	43 mmHg
pO_2	178 mmHg
HCO_3^-	17 mmol/L (17 mEq/L)

After the operation, the patient was successfully extubated. The following results were obtained 24 h after extubation while he was breathing room air:

pH	7.33
pCO_2	36 mmHg
pO_2	82 mmHg
HCO_3^-	19 mmol/L (19 mEq/L)

Case 5

About one month after receiving a liver transplant, a 56-y-o man developed respiratory distress with hyperventilation for several days. On admission, the patient had a 1.5-L pleural effusion that was drained. He was stabilized with an FI-O_2 (%) of 40% while the following blood gas results were obtained:

pH	7.40
pCO_2	28 mmHg
pO_2	126 mmHg
HCO_3^-	18 mmol/L (18 mEq/L)

Case 6

A previously well patient was brought to the emergency room in a moribund state. A chest X-ray showed pulmonary edema. Here are the lab results *(1)*:

pH	7.02
pCO_2	60 mmHg
pO_2	40 mmHg
HCO_3^-	15 mmol/L (15 mEq/L)
Lactate	9.0 mmol/L (81.1 mg/dL)

Case 7

A 60-y-o man, on diuretics for congestive heart failure, has a 5-d history of severe vomiting when he is admitted to the ED. His lab results are as follows:

pH	7.58
pCO_2	35 mmHg

pO_2	95 mmHg
Na	133 mmol/L (133 mEq/L)
K	4.5 mmol/L (4.5 mEq/L)
Cl	80 mmol/L (80 mEq/L)
HCO_3^-	42 mmol/L (36 mEq/L)
Anion gap	11 mmol/L (11 mEq/L)

Case 8

A group of family members came to the ED, all complaining of dizziness and nausea. It was during the winter and none of them is a smoker. These are the lab results for one of the patients:

pH	7.41	
pCO_2	40 mmHg	
pO_2	64 mmHg	
HCO_3^-	24 mmol/L (24 mEq/L)	
sO_2	95%	
Hemoglobin		14.2 g/dL
	$\%O_2Hb$	87.9
	%COHb	6.1
	%metHb	0.8

Case 9

At a routine visit, a diabetic 66-y-o man with a history of COPD has the blood gas and electrolyte results shown below. Ten days later, he is admitted to the ED with diabetic ketoacidosis. Interpret the blood gas and electrolyte results along with the anion gap to assess the patient's acid-base status in each setting. Adapted from *(2)*.

	Units	At routine visit	At admission to ED (10 days later)
pH		7.38	7.22
pCO_2	mmHg	60	58
pO_2	mmHg	78	78
Na	mmol/L	136	138
K	mmol/L	4.0	4.9
Cl	mmol/L	86	87
HCO_3^-	mmol/L	36	26
Anion gap	mmol/L	14	25

CHAPTER 1 DISCUSSION
Acid-Base Exercises

pH	pCO_2 (mmHg)	HCO_3^- (mmol/L)	Duration of medical observation	Acid-base condition
7.40	40	24	—	Very normal acid-base results
7.34	34	18	12 h	Partially compensated metabolic acidosis
7.55	49	40	6 h	Partially compensated metabolic alkalosis
7.55	49	40	24 h	Combined metabolic alkalosis and respiratory alkalosis
7.44	27	18	2 d	Chronic (compensated) respiratory alkalosis
7.23	70	29	6 h	Acute respiratory acidosis
7.23	70	29	2 d	Combined respiratory acidosis and metabolic acidosis
7.28	55	25	2 h	Acute respiratory acidosis
7.41	70	43	2 h	Combined respiratory acidosis and metabolic alkalosis
7.52	31	29	6 h	Combined (mixed) metabolic and respiratory alkalosis
7.08	54	18	12 h	Combined (mixed) metabolic and respiratory acidosis
7.56	20	20	6 h	Acute respiratory alkalosis

Assessment of Cases

Case 1

The patient is in acute ventilatory failure and has acute respiratory acidosis. Bicarbonate has yet to compensate to any significant degree. The patient's oxygenation status is acceptable while breathing 50% oxygen; however, she would likely be hypoxic on room air.

Case 2

On day 10, the patient developed metabolic alkalosis from excessive administration of bicarbonate to correct the metabolic acidosis, as indicated by the elevated lactate. Although the pCO_2 of 46 mmHg might indicate respiratory compensation, the pCO_2 probably reflects inadequate ventilation due to pulmonary edema. The pO_2 of 40 mmHg on 40% oxygen indicates marginal oxygenation status. This patient may have a triple disorder of respiratory acidosis, metabolic acidosis, and metabolic alkalosis.

Revision of the central shunt dramatically improved tissue oxygenation, as determined by the rising pO_2 and decreasing lactate. As the pH normalized, pCO_2 decreased slightly to 44 mmHg, and HCO_3^- decreased to 30 mmol/L (30 mEq/L). The pCO_2 and HCO_3^- at day 15 are more consistent with a compensated respiratory acidosis. These results suggest the remnants of the metabolic alkalosis along with some degree of respiratory compensation.

Case 3

The asthma and several days of hyperventilation suggest this boy has chronic respiratory alkalosis. He is also severely hypoxemic. His electrolyte results show an elevated anion gap. The combination of hypoxemia with an elevated anion gap suggests he is producing excess lactate and has a metabolic acidosis, which could explain why his pH is lower than might be expected for the respiratory alkalosis. He will require supplemental oxygen.

Case 4

During the operation, the patient was in metabolic acidosis. In an alert patient, respiratory compensation would have lowered the pCO_2 more than what was seen here. Therefore, a small component of respiratory acidosis may also be present. After the operation, while breathing room air, the patient is showing a nearly compensated metabolic acidosis. His oxygenation status is acceptable.

Case 5

The patient is in chronic respiratory alkalosis (chronic hyperventilation). Note that renal loss of bicarbonate has fully compensated and brought the pH to normal, as can happen in chronic respiratory alkalosis.

Case 6

This patient has acute ventilatory failure (respiratory acidosis) from the pulmonary edema combined with metabolic acidosis from moderate to severe hypoxemia. The poor oxygenation in the lungs likely explains the decreased pO_2 of 40 mmHg and the congestive heart failure could cause poor perfusion, both of which could cause the elevated lactate. The very low pH reflects the combined respiratory and metabolic acidoses.

Case 7

This case illustrates the importance of the clinical history in accurate interpretation of the case. The chronic use of diuretics could cause a compensated metabolic alkalosis. From that, we would expect an elevated bicarbonate and pCO_2 with a normalized pH. However, the recent vomiting would also cause a metabolic alkalosis, which would elevate the pH and bicarbonate even further. This patient has a metabolic alkalosis superimposed on a chronic metabolic alkalosis, which is a mixed acid-base disorder.

Case 8

The clues to this case are that both dizziness and nausea are occurring in several family members and that it is during the winter. This suggests exposure to a source of carbon monoxide, such as a heater. The moderately elevated %CO Hb is consistent with this. The lab results also show the difference between the sO_2 (nearly normal) and the $\%O_2Hb$ (clearly decreased), which is typical for CO exposure. It is likely that this person's %COHb was higher when he left his house and has decreased somewhat during the trip to the ED. In fact, their heater was faulty and putting out CO.

Case 9

At the routine visit, the elevated pCO_2 and slightly decreased pO_2 are consistent with the COPD. Based on the normal pH with elevated HCO_3^-, he appears to have a chronic respiratory acidosis with metabolic (renal) compensation elevating his HCO_3^-. The anion gap (AG) is normal. When he is admitted to the ED 10 days later, the patient has developed a metabolic acidosis with an elevated AG. Because we know the prior results, the ED results are easier to interpret. If we had only the ED results with no history, the patient might appear to have an acute respiratory acidosis. However, the elevated AG suggests a metabolic acidosis and the bicarbonate would be expected to be lower. In fact, the patient has developed a recent metabolic ketoacidosis, but because of the chronic respiratory acidosis with compensation (by metabolic alkalosis), the HCO_3^- is normal.

REFERENCES

1. Iberti TJ, Leibowitz AB, Papadakos PJ, Fischer EP. Low sensitivity of the anion gap as a screen to detect hyperlactatemia in critically ill patients. Crit Care Med 1990;18:275–7.
2. University of Connecticut. Acid base online tutorial. http://fitsweb.uchc.edu/student/selectives/TimurGraham/Case_6.html (Accessed Dec 2008).

CHAPTER 2

Questions

1. Which clinical findings are typical of hyperparathyroidism?
   a. hyperphospatemia
   b. hypophospatemia
   c. hypocalcemia
   d. hypercalcemia
   e. b and d

2. Which of the following is NOT a cause of hypercalcemia?
   a. hyperparathyroidism
   b. malignancy
   c. vitamin D deficiency
   d. drugs
   e. endocrine disorders

3. Which clinical condition is most commonly associated with hypermagnesemia?
   a. increased dietary intake
   b. hypoaldosteronism
   c. metabolic acidosis
   d. renal failure
   e. hypokalemia

4. Which statement is NOT correct about phosphorous homeostasis?
   a. Inorganic phosphate is freely filtered by the glomerulus.
   b. All phosphate transport is by active mechanisms.
   c. PTH increases phosphate loss in urine.
   d. Vitamin D increases intestinal absorption and renal reabsorption of phosphate.
   e. Administration of growth hormone increases phosphate levels in the blood.

5. True or False. Most phosphorous in the blood is organic phosphates.

6. Which of the following clinical conditions can cause hyperphosphatemia?
   a. both acute and chronic renal failure
   b. leukemia
   c. rhabdomyolysis
   d. decreased glomerular filtration rate
   e. all the above

7. Hypophosphatemia is NOT found in which condition?
   a. alcoholism
   b. GI maladsorption conditions

c. respiratory alkalosis
   d. chronic renal failure
   e. excess insulin administration

8. Hypocalcemia is NOT associated with which of the following conditions?
   a. following surgery in the neck area
   b. hypoparathyroidism
   c. inadequate intake of calcium and vitamin D
   d. the neonatal period
   e. prolonged immobilization

9. A patient has renal failure and elevated serum phosphate and PTH levels and a chronically low serum Ca. What is the most likely diagnosis from the following choices?
   a. primary hyperparathyroidism
   b. secondary hyperparathyroidism
   c. a PTH-producing tumor
   d. tertiary hyperparathyroidism
   e. hypomagnesemia

10. The patient mentioned above (question 9) has a renal transplant. Three months later, the patient's plasma calcium level is consistently elevated. What is the most likely cause?
    a. primary hyperparathyroidism
    b. secondary hyperparathyroidism
    c. a PTH-producing tumor
    d. tertiary hyperparathyroidism
    e. hypomagnesemia

Cases:

Assess the clinical status of the following patients and their laboratory results for possible disorders of calcium and/or magnesium metabolism.

Case 1

A 24-y-o woman was admitted for treatment of systemic lupus erythematosus. She complained of a weight gain of ~2.3 kg during the past week and a bloated feeling. She was started on the diuretic furosemide, 20 mg every 4 h. Twenty-four hours later the patient complained of numbness and tingling in her face, hands, and feet, which rapidly progressed to acute, painful carpal spasm. Lab results at this time were as follows:

Total calcium	2.05 mmol/L (8.2 mg/dL)
Ionized calcium	1.10 mmol/L (4.4 mg/dL)
Magnesium	0.50 mmol/L (1.2 mg/dL)

The patient was treated with 1 g of $MgSO_4$ and 0.1 g of calcium gluconate given by iv over 10 min. Within 60 min, the patient was free of all numbness, tingling, and spasm. Lab results 2 h later:

Total calcium	2.50 mmol/L (10.0 mg/dL)
Ionized calcium	1.30 mmol/L (5.2 mg/dL)
Magnesium	2.50 mmol/L (6.1 mg/dL)

(Adapted from a case history provided by Ronald J. Elin, Chief, Clinical Pathology Department, Clinical Center, National Institutes of Health, Bethesda, MD.)

Cases 2–5

Evaluate the following lab results for causes of hyper- and hypocalcemia. Consider the following possible diagnoses: hyperparathyroidism, malignancy, chronic renal disease, and post surgery in the neck.

Lab Test	Reference range	Case 2	Case 3	Case 4	Case 5
Midregion PTH (µg/L)	0.40–1.50	0.10	116	0.90	3.80
Intact PTH (ng/L)	13–64	2	951	3	65
Total Ca					
(mmol/L)	2.10–2.55	1.88	1.98	4.10	2.65
(mg/dL)	8.41–10.2	7.5	7.9	16.4	11.8
Ionized Ca (mmol/L)	1.16–1.32	1.11	1.07	2.34	1.46
Phosphate					
(mmol/L)	0.87–1.49	1.19	2.07	1.19	0.90
(mg/dL)	2.7–4.6	3.7	6.4	3.7	2.8

Case 6

A 52-y-o man was admitted for abdominal pain and vomiting. He had a normal weight, and no history of surgery, GI disorders, or significant alcohol use. On admission, his amylase, lipase, and other serum enzymes (AST, ALT, ALP) were elevated and his phosphate and calcium were slightly decreased. His blood gases indicated a degree of respiratory alkalosis (pH 7.49; pCO_2 30 mmHg). An ultrasound showed gallstones and an enlarged pancreas. After three days, his serum phosphate became very low at 0.9 mg/dL. Intravenous phosphate was given during the next few days until the hypophospatemia resolved. What condition do the clinical and laboratory findings indicate and how does this relate to the phosphate level (1)?

Case 7

A diabetic patient on insulin has a plasma phosphorous of 0.60 mmol/L (1.9 mg/dL) and a decreased phosphorous excretion of 70 mg/ 24 h. What is the most likely explanation for these results?

Appendix

a. hyperparathyroidism
b. hypoparathyroidism
c. overdose of insulin
d. vitamin D deficiency
e. dietary deficiency of phosphorous

CHAPTER 2 DISCUSSION
Answers to Questions

1. e.
2. c.
3. d.
4. b.
5. True
6. e.
7. d.
8. e.
9. b.
10. d.

Assessment of Cases

Case 1

The patient experienced an acute rapid change in serum magnesium concentration, which caused more severe clinical symptoms than seen with a more gradual change. The hypomagnesemia was accompanied by a mild hypocalcemia. The cause of the rapid decrease in magnesium was the repeated doses of furosemide. Furosemide is a diuretic that acts primarily on the ascending loop of Henle to increase urinary loss of magnesium. Administration by iv of Mg and Ca relieved the symptoms of numbness and tingling. The elevated magnesium after treatment should quickly decline with renal excretion.

Case 2

These results are from a patient 24 h after parathyroid surgery for correction of hyperplasia of the parathyroid glands. Frequently, parathyroid function is impaired for several days after surgery, which results in a temporary condition of hypoparathyroidism. The decreased PTH and calcium results are typical of hypoparathyroidism. (Note that PO_4 is not always elevated in hypoparathyroidism.)

Case 3

This is a patient with chronic renal failure who has developed secondary hyperparathyroidism with renal osteodystrophy. Note the extremely elevated midregion PTH (77 times the upper reference range) relative to intact PTH (15 times the upper reference range). Renal failure leads to hyperphosphatemia and abnormal vitamin D and/or PTH metabolism. The chronic hyperphosphatemia and hypocalcemia can sometimes result in extremely high concentrations of PTH, which is appropriately secreted in an attempt to increase calcium and lower phosphate. Unfortunately, this chronic condition of HPTH often causes a spectrum of bone diseases, called renal osteodystrophy.

Case 4

These results are typical of a patient with hypercalcemia of malignancy, in which extreme hypercalcemia develops. A non-PTH hypercalcemic factor is likely being produced by the malignant tissue and secreted into blood. The intact PTH result indicates that the parathyroid glands have appropriately diminished production of intact PTH. Midmolecule PTH does not show the diminished parathyroid response due to the slower clearance from blood.

Case 5

This is a patient with results typical of primary HPTH: moderate elevations of calcium and PTH with slightly decreased phosphate. Although intact PTH is only slightly elevated, the midregion PTH assay is clearly elevated, because it detects many of the fragments from intact PTH. Ionized calcium is elevated slightly more (11%) than is total calcium (4%). PTH promotes renal loss of phosphate, which explains the low normal serum phosphate observed in this patient.

Case 6

This patient clearly has pancreatitis, consistent with many of his laboratory results. These include the elevated amylase and lipase, the respiratory alkalosis, and the decreased calcium levels. While hypophosphatemia has not frequently been found in pancreatitis, a recent report suggests hypophosphatemia may often have been missed (1). Certainly, the massive inflammatory response of pancreatitis, similar to sepsis, can release many factors that can cause electrolyte abnormalities, including hypophosphatemia.

Case 7

Since the administration of insulin causes a shift of phosphate from extracellular to intracellular compartments, the most likely cause of the decreased plasma and urine phosphorous in this diabetic patient would be excess administration of insulin. Hyperparathyroidism can lower plasma phosphorous, but should increase urinary phosphorous. A chronic dietary deficiency of phosphorous could also cause these findings, but there is no evidence for this situation occurring.

REFERENCE

1. Steckman DA, Marks JL, Liang MK. Severe hypophosphatemia associated with gallstone pancreatitis: a case report and review of the literature. Dig Dis Sci 2006;51:926–30.

Appendix

CHAPTER 3

Cases:

Assess the clinical status of the following patients and their laboratory results for possible disorders of water and/or electrolyte balance.

Case 1

A 56-y-o man diagnosed with an oat-cell carcinoma of the lung develops progressive lethargy and confusion. The following laboratory data were obtained:

Na	119 mmol/L (119 mEq/L)
Glucose	6.22 mmol/L (112 mg/dL)
K	4.6 mmol/L (4.6 mEq/L)
BUN	3.2 mmol/L (9 mg/dL)
Cl	77 mmol/L (77 mEq/L)
Serum osmolality	251 mmol/L (251 mOsm/kg)
HCO_3^-	26 mmol/L (26 mEq/L)
Urine osmolality	857 mmol/L (857 mOsm/kg)

What is the most likely cause of this patient's abnormal sodium? Is the calculated osmolality consistent with the measured osmolality?

Case 2

A 21-y-o woman with insulin-dependent diabetes was brought to the hospital in a coma. On admission she was hypotensive (blood pressure 90/20 mmHg) with a rapid pulse (122/min) and rapid breathing (32/min). Her laboratory data were as follows:

Na	134 mmol/L (134 mEq/L)
K	6.4 mmol/L (6.4 mEq/L)
Glucose	66.1 mmol/L (1200 mg/dL)
pH	6.80
pCO_2	10 mmHg
HCO_3^-	3 mmol/L (3 mEq/L)

What is the likely cause of these abnormalities? What two therapeutic agents would you expect to be given? What would be the effect of this therapy?

Case 3

A 40-y-o woman with edema is found on exam to be hypertensive, with the following laboratory data:

Na	145 mmol/L (145 mEq/L)
K	2.8 mmol/L (2.8 mEq/L)
Cl	106 mmol/L (106 mEq/L)

HCO_3^-	30 mmol/L (30 mEq/L)
Arterial pH	7.48
Urine K	50 mmol/L (50 mEq/L) (abnormally high for hypokalemia)

What does the combination of elevated sodium with decreased potassium suggest in this patient?

Case 4

A 62-y-o man with congestive heart failure and mild renal failure was treated with the angiotensin-converting enzyme inhibitor captopril to relieve the congestive heart failure. The following lab results were obtained:

Na	134 mmol/L (134 mEq/L)
K	6.8 mmol/L (6.8 mEq/L)
Cl	102 mmol/L 102 mEq/L)
HCO_3^-	21 mmol/L (21 mEq/L)
BUN	10.0 mmol/L (28 mg/dL)
Creatinine	159.12 µmol/L (1.8 mg/dL)
Glucose	6.38 mmol/L (115 mg/dL)

After further treatment with sodium polystyrene sulfonate, plasma K decreased to 4.8 mmol/L (4.8 mEq/L).

What caused the initial elevated potassium and the later normalized potassium?

Case 5

An elderly man was evaluated for having nausea and vomiting for four weeks. He was taking no drugs and had no signs of hypovolemia, had a normal blood pressure and pulse, and his skin turgor appeared normal. His renal function was also normal. His lab results on plasma were:

Na	125 mmol/L
K	4.4 mmol/L
Cl	97 mmol/L
Total CO_2	20 mmol/L
Glucose	75 mg/dL
Creatinine	0.9 mg/dL
pH	7.44

Aldosterone and cortisol were both very low.

Evaluate this patient for possible explanations of his low Na, normal K, and low aldosterone.

Case 6

A serum K level is ordered in a patient with thrombocytosis and a platelet count of 1×10^6 / mm^2. The serum K level is 6.3 mmol/L. Considering both

Appendix   113

clinical and pre-analytical possibilities, what would be the next logical action for the physician to take?

a. Give the patient calcium gluconate.
b. Administer glucose and insulin.
c. Draw another blood sample for a serum K level.
d. Draw a blood sample for a plasma K level.
e. Put the patient on renal dialysis.

CHAPTER 3 DISCUSSION

Case 1

This patient had the classic symptoms of the syndrome of inappropriate secretion of ADH (SIADH): decreased serum sodium, decreased serum osmolality, and increased urine osmolality. The lower-normal urea N is also typical of SIADH.

Oat-cell carcinomas often produce a peptide with ADH-like activity. Usually, acute therapy for SIADH is simply to restrict water intake. However, because of the more severe symptoms indicating cerebral edema, the patient was treated with hypertonic saline and furosemide (a diuretic). The diuretic was given to prevent hypervolemia.

The calculated osmolality is 247 mOsm/kg:

$$(2 \times 119) + 3.2 + 6.22 = 247$$

which is close to the measured osmolality of 251 mmol/kg.

Case 2

This patient was in severe diabetic ketoacidosis. She was given 8 U of insulin immediately and 8 U/h thereafter. She was also given iv injections of sodium bicarbonate. The insulin decreased glucose at ~5.55 mmol/L (100 mg/dL) per hour. Insulin will also lower the plasma potassium. Sodium bicarbonate will increase the pH and provide much-needed additional buffer capacity because the patient is nearly depleted. Increasing the pH will also lower the potassium concentration (case adapted from ref. *1*).

Case 3

The plasma value for sodium and potassium suggests mineralocorticoid excess, most likely caused by primary hyperaldosteronism. The combination of hypertension, hypokalemia, metabolic alkalosis, and inappropriately increased urinary potassium (urinary K should be decreased with hypokalemia) suggests a possible diagnosis of primary hyperaldosteronism (case adapted from ref. *2*).

Case 4

Captopril is a vasodilator that diminishes angiotensin by inhibiting angiotensin-converting enzyme. Although captopril is effective in the treatment of congestive heart failure, hyperkalemia may result possibly by the inhibition of aldosterone secretion. Sodium polystyrene sulfonate is an ion-exchange resin that lowers plasma K by binding to K in the gut. In this case, it brought the plasma K down to 4.8 mmol/L (4.8 mEq/L).

Case 5

This patient has two disorders that could affect the K and other electrolytes in opposite directions. The low aldosterone should cause a low Na, an elevated K and Cl, and possibly a mild acidosis. His long history of vomiting should cause an alkalosis, with an elevated total CO_2, a decreased K, and a decreased Cl. The best explanation is that this person has multiple disorders that have offset each other and made the interpretation more difficult. These two conditions have combined to produce hyponatremia, with a normal K, and nearly normal Cl, bicarbonate, and pH. The very low cortisol and aldosterone suggest Addison's disease. As mentioned in the text, while an elevated K is associated with Addison's disease, about one-third of cases have a normal plasma K.

Case 6

While actions a, b, and e will lower a blood K level, both a and e are for more drastic hyperkalemia and b may not be necessary. Since a serum K was ordered, the high platelet count suggests that the clotting process may have released some K into the serum and caused the modest hyperkalemia. Thus, a serum K would likely be a waste of time; getting a plasma K should prove that the patient's K level is actually normal, or (if plasma K still elevated) would confirm the hyperkalemia. So answer **d** is correct.

REFERENCES

1. Wahr JA, Parks R, Biosvert D, et al. Preoperative serum potassium levels and perioperative outcomes in cardiac surgery patients. JAMA 1999;281:2203–10.
2. Rose BD. Clinical physiology of acid-base and electrolyte disorders, 3rd ed. New York: McGraw-Hill, 1989.

Index

acetazolamide, 71, 96
acid-base status
  anion gap changes in, 6
  bicarbonate in, 96
  clinical abnormalities of, 15–16
  clinical conditions related to, 27
  evaluation of, 26–31
  nomogram, 30
  regulation of, 11–13
  respiratory component of, 8
  sample collection, 35–36
  sample handling, 35–36
acidemia, 14, 15
acidity, pH and, 2. *See also* pH
acidosis. *See also* metabolic (nonrespiratory) acidosis; respiratory acidosis
  hypophosphatemia and, 68
  oxygen deficits and, 9
acids
  definition of, 2 (*See also* pH)
  physiology of, 8–9
adrenal insufficiency, 51, 85
albumin. *See also* hypoalbuminemia
  buffering by, 6, 11
  calcium binding to, 45–46, 50
  plasma levels, 19
  renal diseases and, 47
alcoholism
  hypophosphatemia and, 68
  magnesium deficiency in, 58, 62
aldosterone
  blood volume changes and, 80
  chloride regulation and, 95, 96
  function of, 79
alkalemia
  2,3-DPG and, 14
  hypokalemia and, 90–91
  potassium levels and, 90
alkali intake, bicarbonate levels and, 97
alkalinity, pH and, 2. *See also* pH
alveolar-arterial oxygen gradient, 33, 34
American Diabetes Association, 61–62
anemia, 15, 33
angiotensin-converting enzyme, 80
angiotensin II, 79, 80
anion gap (AG)
  acid-base disorders and, 31
  causes of increase in, 16
  changes in acid-base disorders, 6
  description of, 5
  reference range, 3, 81
antacids
  hypercalcemia and, 51
  hypermagnesemia and, 64
  hypophosphatemia and, 68, 70
antibiotics, magnesium levels and, 60
anticoagulants
  artifactual hyperkalemia and, 94
  blood gas samples and, 35, 36
  calcium samples and, 64
  magnesium samples and, 64
antidiuretic hormone (ADH)
  chloride regulation and, 95
  deficiency, 88
  half-life of, 78
  hyponatremia and, 82
  osmolality and, 78
  release of, 85
  secretion of, 80, 86
Aoyagi, Takuo, 2
apgar scores, 36

Index

Arrhenius, 2
arterial blood, reference ranges, 3

base excess (BE), 3, 4, 5
bases, physiology of, 9. *See also* pH
bicarbonate (HCO_3^-). *See* HCO_3^- (bicarbonate)
bicarbonate–carbon dioxide (carbonic acid), 9–10
blood
  arterial, 3, 35
  color changes, 13
  magnesium regulation in, 55–56
  oxygen content, 32, 33
  oxyhemoglobin dissociation, 13
  sample collection, 35–36
  sample handling, 35–36
  storage, 36
  storage temperatures, 36
  transfusions, 92, 93
  venous, 3, 35
  volume of, 79–81
blood flow, to tissues, 15
blood gases
  cord blood, 36–37
  evaluation of results, 26–31
  history of measurement, 1–2
  reference ranges, 26
  sample collection, 35–36
  sample handling, 35–36
bone
  calcium absorption, 44
  hungry bone syndrome, 49
  magnesium release from, 55
bronchodilators, 90
buffer systems, 9–11, 96
burn injuries
  hypernatremia and, 87
  hyperphosphatemia and, 71
  hyponatremia and, 82
  hypophosphatemia and, 68
  magnesium deficiency and, 63

caffeine, hypokalemia and, 90
calcitonin, 44–45
calcium. *See also* hypercalcemia; hypocalcemia distribution, 45–46
  function of, 41
  hypercalcemia, 42

  interpretation of measurements, 52–53
  ion channels, 42
  measurements of, 46
  neonatal monitoring, 48–49
  physiology, 42–43
  PTH and, 53
  reference ranges, 54
  regulation in blood, 43–45
  renal diseases and, 47
  sample collection, 53–54
  sample handling, 53–54
carbon dioxide, 81. *See also* pCO_2
carbonic acid (bicarbonate–carbon dioxide), 9–10
carbonic anhydrase, 96
carboxyhemoglobin (CI-Hb), 8
cardiac function. *See also* myocardial infarction
  calcium and, 42
  hyperkalemia and, 93–94
  hypokalemia and, 91
  low magnesium and, 60
  magnesium and, 59–60
catecholamines, 90
Cavendish, Henry "Hank," 1
chemotherapy, electrolyte balance in, 71, 85, 92
chloride (Cl). *See also* hyperchloremia; hypochloremia acid-base disorders and, 30–31
  hyperchloremia, 30, 95–96
  hypochloremia, 30–31, 95–96
  physiology, 95
  reference ranges, 81
  regulation, 95
  strong ion difference and, 6–7, 95
cirrhosis, hyponatremia and, 82, 83, 86
cisplatin, 60, 61
citrate, sample collection and, 45, 64
Clark, Leland, 2
compensation
  in acid-base regulation, 12–13, 18
  calculations for, 17–18
  HCO_3^- in, 17
  in metabolic acidosis, 16–17, 24
  in metabolic alkalosis, 19–20, 24
  pCO_2 in, 17
  pH in, 17
  in primary disorders, 23, 29–30

Index

in respiratory acidosis, 20–21, 24
in respiratory alkalosis, 22–23, 24
cord blood
  blood gases, 36–37
  reference ranges, 37
corticosteroids, 19
cortisone, hyperchloremia and, 96
creatinine
  acid-base disorders and, 31
  elevated, 69–70
  hyponatremia and, 84
critical illness
  hypocalcemia in, 50
  hypomagnesemia in, 58–59
Cushing's syndrome, 19, 88
cyclosporine, magnesium and, 60, 61

dehydration thirst and, 79
delta ratios, 25, 31
desmopressin, hyponatremia and, 85
diabetes insipidus, 79
diabetes mellitus, 61–62, 93
diabetic ketoacidosis. *See* ketoacidosis
diarrhea
  anion gap changes in, 6
  bicarbonate levels and, 97
  hypokalemia and, 90
  hypophosphatemia and, 68
  magnesium deficiency in, 62
  sodium levels and, 84
digoxin, magnesium levels and, 61
2,3-diphosphoglycerate (2,3-DPG), 14, 69
diuretics, 19
  chloride loss and, 95
  effect on magnesium levels, 60
  hypokalemia and, 90
  hypophosphatemia and, 68
DO_2 (oxygen delivery), 8

eclampsia of pregnancy, 62
elderly patients
  altered mental status in, 87
  dehydration in, 79
  diminished thirst reflex in, 87
  hyponatremia in, 82, 86
electrolytes. *See also specific* electrolytes
  diagnostic tests, 41–76
  history, 41–42
  reference ranges, 81
  significance of, 41–42

endocrine disorders, 71
enzyme function, magnesium and, 54
epinephrine, 42, 80, 90
Epsom salts, 41
ethylenediamenetetraacetic acid
  (EDTA), 64
euvolemic status, 84

familial hypocalciuric hypercalcemia
  (FHH), 51
fetuses, acid-base balance, 36–37
FI-O_2, pO_2 and, 32
Franklin, Benjamin, 1

gas exchange, 14–15
gastric fluid content, 7
gastric suction, 90, 95
gentamicin, 60, 61
glomerular filtration rate (GFR)
  blood volume changes and, 81, 84
  hyperkalemia and, 93
  hyperphosphatemia and, 70, 71
  PTH levels and, 53
glucose, phosphate shifts and, 68
Grew, Nehemiah, 41
Grogono, Alan, 1

Hasselbalch, Karl, 2
Hastings, Baird, 41
HCO_3^- (bicarbonate)
  in acid-base disorders, 18
  base excess compared to, 5
  calcium binding to, 45
  changes in respiratory
    conditions, 24
  chloride regulation and, 95
  in compensation, 17
  decreased, 29, 97
  description, 4
  evaluation of, 26
  excess, 19
  hypokalemia and, 91
  increased, 29, 97
  loss of, 15
  in metabolic alkalosis, 19
  in mixed disorders, 17
  normal, 29
  pCO_2 relationship with, 18
  physiology, 96
  in primary disorders, 17

HCO_3^- *(continued)*
  production of, 9
  reference ranges for, 3
  regulation, 96
heart failure, hyponatremia and, 82
hemoglobin, reduced (HHb), 11
hemoglobin (Hb)
  binding to oxygen, 13–14
  buffering by, 10
  fetal, 13–14
  measurement of, 7
  oxygen saturation of, 32
  S-nitroso (SNO-Hb), 14
  structure of, 13
hemolysis
  artifactual hyperkalemia and, 94
  hyperkalemia and, 92
  hyperphosphatemia and, 69, 71
Henderson, Lawrence, 2
heparin
  calcium binding by, 53–54
  effect on magnesium
    specmens, 64
  liquid, 36
hungry bone syndrome, 49
hydrogen (H^+), 7. *See also* pH
hyperaldosteronism, primary, 88
hypercalcemia. *See also* calcium
  causes of, 50–52
  evaluation of, 51
  hormonal responses to, 43
  malignancies and, 42, 45, 46
  parathyroid disease and, 46
  symptoms of, 50
hyperchloremia. *See also* chloride (Cl)
  acid-base disorders and, 30
  causes of, 95–96
hyperkalemia. *See also* potassium (K)
  artifactual, 94
  causes of, 92–93, 95–96
  clinical effects of, 93–94
  sample collection, 94
  sample handling, 94
  treatment of, 94
hypermagnesemia, 63–64. *See also* magnesium
hypernatremia. *See also* sodium (Na)
  evaluation of, 86–88
  nonrenal causes, 87
  renal causes, 88
  thirst and, 79
  treatment of, 88
hyperosmolality, 92, 93. *See also* osmolality
hyperoxemia, 25
hyperparathyroidism. *See also* parathyroid hormone (PTH)
  hyperchloremia and, 96
  hypophosphatemia and, 68
  malignancies and, 52
hyperparathyroidism (HPTH), 50–52
hyperphosphatemia. *See also* phosphate
  calcium levels and, 46
  causes of, 70–71
  clinical consequences of, 71
  evaluation of, 69–70, 70–71
  metabolic acidosis and, 69–70
  renal diseases and, 47
  treatment of, 71
hyperthyroidism, 51
hyperventilation, 11, 22
hypervolemia, 84, 86
hypoalbuminemia, 11. *See also* albumin
hypocalcemia. *See also* calcium
  causes of, 46–50, 47
  evaluation of, 48
  hormonal responses to, 43
  hungry bone syndrome and, 49
  in neonates, 48–49
  renal diseases and, 47
  symptoms of, 43
hypochloremia. *See also* chloride (Cl)
  acid-base disorders and, 30–31
  causes of, 95–96
hypodipsia, 87. *See also* thirst
hypokalemia, 19. *See also* potassium (K)
  acid-base disorders and, 30
  bicarbonate levels and, 97
  causes of, 90–91
  clinical effects of, 91
  treatment of, 91
hypomagnesemia. *See also* magnesium
  causes of, 57–63
  definition of, 57
  effects of, 55
  hypocalcemia in, 50
  hypokalemia and, 90
  hypophosphatemia and, 68
  insulin resistance and, 61

postoperative, 60
symptoms of, 57
treatment of, 63
hyponatremia. *See also* sodium (Na)
  ADH release and, 85
  artifactual, 85
  differential diagnoses, 83
  evaluation of, 82–86
  hypervolemia and, 86
  laboratory tests, 83
  patient history in, 83
  treatment of, 86
  volume status-related, 84
hypoparathyroidism, 49
hypophosphatemia. *See also* phosphate
  causes of, 68
  evaluation of, 66
  symptoms of, 69
  treatment of, 69
hypothermia, 90, 92
hypoventilation, 11, 19
hypovolemic status, 84
hypoxemia, 32
hypoxia
  anion gap and, 16
  apgar scores and, 36
  base excess and, 4, 97
  cellular, 49, 58, 62
  cerebral edema and, 82
  hyperkalemia and, 92
  hyperventilation and, 27
  metabolic acidosis and, 5, 97
  myocardial, 60
  tissue, 25, 92

insulin
  deficiency, 92
  effect on magnesium concentration, 55
  hypokalemia and, 90, 91
  phosphate shifts and, 68
  potassium levels and, 90
insulin resistance, 61
intensive care units, monitoring in, 49
intestinal fistulae, 90
intrapulmonary shunting, 15, 33–34
ion-sensitive electrodes, 41
iron, hemoglobin, 13
irritability, hypokalemia and, 91
isoproterenol, 42

ketoacidosis
  anion gap changes in, 6
  bicarbonate levels and, 97
  description, 9
  hypokalemia and, 90
  hypophosphatemia and, 68
kidneys. *See also* renal diseases
  acid-base regulation and, 12
  acute renal failure, 51
  diseases of, 47
  electrolyte regulation by, 41
  hyperkalemia and, 93
  magnesium regulation by, 55
  phosphate regulation by, 66
  potassium regulation by, 89–90
  renal tubular acidosis, 6
  role of PTH, 44
  sodium excretion, 81
  sodium regulation by, 86
Kramer, Kurt, 2

lactate
  acid-base disorders and, 31
  calcium binding to, 45
  strong ion differences, 7
lactate acidosis, 8–9
  anion gap changes in, 6
laxative use, 90
leukemia, 71
leukocytosis, 36
lungs, gas exchange in, 14–15

magnesium, 54–65. *See also* hypermagnesemia; hypomagnesemia
  dietary recommendations, 55
  distribution, 56–57
  in energy metabolism, 59–60
  evaluation of status, 57
  function of, 41
  GI absorption of, 55
  load tests, 57
  physiology, 54
  protein binding of, 57
  reference ranges for, 65
  regulation in blood, 55–56
  renal diseases and, 47
  replacement therapy, 58
  retention test, 64–65
  sample collection, 64

magnesium (*continued*)
  sample handling, 64
  urinary levels, 57
magnesium deficiency. *See also*
    hypomagnesemia
  calcium levels and, 46
  diagnosis of, 64–65
  dietary, 62
malignancies
  calcium levels in, 46
  hungry bone syndrome and, 49
  hypercalcemia in, 42, 45, 50
  hyperparathyroidism from, 52
  hyperphosphatemia and, 71
malignant hyperthermia, 71
McLean, Franklin, 41
metabolic acid, 8
metabolic (nonrespiratory) acidosis, 15–17
  causes of, 15–16
  compensation in, 16–17, 24
  delta ratios in, 25
  HCO_3^- in, 17
  hyperkalemia and, 92, 93
  hyperphosphatemia and, 69–70
  pCO_2 in, 17
  pH in, 17
  treatment of, 18
  use of the anion gap in, 16
metabolic (nonrespiratory) alkalosis, 18–20
  causes of, 18–19
  chloride loss and, 95
  compensation in, 24
  HCO_3^- in, 17
  pCO_2 in, 17
  pH in, 17
metabolic (renal) system, 12
metabolic status, evaluation of, 28–29
milk alkali syndrome, 51
Millikan, Glen, 2
mixed disorders
  anion gap changes in, 6
  detection of, 23
  evaluation for, 29–31
  HCO_3^- in, 17
  pCO_2 in, 17
muscle weakness, 91, 93–94
myocardial infarction, 60

natriuretic peptides, 80
neonates
  hypernatremia in, 87
  monitoring calcium levels in, 46, 48–49
neuromusclar irritability, 46
nitric oxide (NO), 14, 59
norepinephrine, 80

%O_2Hb, 3, 7–8
osmolality. *See also* hyperosmolality
  blood volume and, 79–81
  calculation of, 78
  normal, 78
  physiology of, 77–78
  reference ranges, 81
  regulation of, 78–79
  volume regulation and, 77–78
  water deficits and, 79
  water load and, 79
oxalate, specimen collection and, 64
oximeters, 1
oximetry, definition of, 1
oxygen. *See also* pO_2; sO_2
  deficits, 9
  delivery to tissues, 14–15
  hemoglobin binding to, 13–14
  release to tissues, 15
oxygen consumption (VO_2), 8
oxygenation
  arterial, 31–35
  disorders of, 25
  DO_2 (oxygen delivery), 8
  evaluation of, 31–35
  failure of, 34
  tissue, 34–35
oxyhemoglobin dissociation curve, 13
oxyntic cells, 6, 7

Paget's disease, 63
pancreatitis, 46, 50
paO_2, 34
paralysis, hypokalemia and, 91
parathyroid gland
  dysfunction of, 42, 46
  hypoparathyroidism, 49
parathyroid hormone (PTH)
  deficiencies, 46, 63
  effect of hypophosphatemia on, 69

interpretation of measurements, 52–53
in regulation of calcium, 43–45
in regulation of electrolytes, 41
in regulation of magnesium, 55
renal diseases and, 47
synthesis of, 44
total calcium and, 53
parathyroid hormone-related peptide (PTHrP), 43
pCO_2. *See also* carbon dioxide
in acid-base disorders, 18
changes in respiratory conditions, 24
in compensation, 17
cord blood, 37
decreased, 28
description, 3
effect of air bubbles on, 35
in evaluating oxygenation, 34
evaluation of, 26
HCO_3^- relationship with, 18
in HCO_3^--CO_2 buffer system, 11
increased, 28
lung ventilation and, 8
in mixed disorders, 17
normal, 28
in primary disorders, 17
reference ranges for, 3
in respiratory acidosis, 20–21
pH, 2
acid-base disorders and, 30
acid-base status and, 30
changes in respiratory conditions, 24
compensation response, 12–13, 17
of cord blood, 37
evaluation of, 26, 27–28
in metabolic acidosis, 15
in metabolic alkalosis, 19
in mixed disorders, 17
potassium levels and, 89
in primary disorders, 17
reference ranges for, 3
renal diseases and, 47
in respiratory acidosis, 20–21
strong ion difference and, 5–7
phosphate, 66–71. *See also* hyperphosphatemia; hypophosphatemia
buffering by, 11
calcium binding to, 45
depletion of, 41

distribution of, 66–67
GI absorption of, 66
physiology, 66
reference ranges for, 71–72
regulation in blood, 66
renal diseases and, 47
plasma
osmolality of, 77, 78
sodium chloride in, 78
pO_2. *See also* oxygen
arterial, 27, 35
arterial oxygenation and, 31–35
in cord blood, 37
description, 3
effect of air bubbles on, 35
evaluation of, 26
gas exchange and, 14–15
reference ranges for, 3
polydipsia, 85
potassium (K). *See also* hyperkalemia; hypokalemia
acid-base disorders and, 30
physiology of, 88–89
reference ranges, 81
regulation of, 89–90
sodium levels and, 84
strong ion differences, 7
pregnancy
eclampsia of, 62
hyponatremia and, 85
Priestley, Joseph, 1
primary disorders
compensation in, 23, 29–30
pCO_2 in, 17
proteins, buffering by, 11
pseudohypoparathyroidism, 49
psychoactive agents, 85

renal diseases, 47–48. *See also* kidneys
calcium levels in, 46
hyperkalemia and, 93
hypermagnesemia in, 63–64
hyperphosphatemia and, 70
hypokalemia and, 90
renal failure, acute, 51
renal tubular acidosis, 6
renin, secretion of, 79
renin-angiotensin-aldosterone system, 80
respiratory acidosis, 20–21
causes of, 11–12, 20

respiratory acidosis (*continued*)
  chloride loss and, 96
  compensation in, 20–21, 24–25
  HCO_3^- in, 17
  pCO_2 in, 17
  pH in, 17
  treatment of, 21
respiratory alkalosis
  causes of, 22
  compensation in, 22–25
  description, 22
  HCO_3^- in, 17
  pCO_2 in, 17
  pH in, 17
  treatment of, 23
respiratory (ventilatory) system, 11–12
rhabdomyolysis, 71
Ringer, Sydney, 41

S-nitroso hemoglobin (SNO-Hb), 14
salt-wasting nephropathy, 84
sample collection
  for blood gas determination, 35–36
  for calcium measurements, 53–54
  in hyperkalemia, 94
  for magnesium evaluation, 64
sample handling
  artifactual hyperkalemia and, 94
  for blood gas determination, 35–36
  for calcium measurements, 53–54
  in hyperkalemia, 94
  for magnesium evaluation, 64
sarcoidosis, 51
serum, osmolality of, 77
shunting, intrapulmonary, 15, 33–34
silicone, specimen collection and, 64
Simon, Wilhelm, 65
SNO-Hb (S-nitroso hemoglobin), 14
sO_2. *See also* oxygen
  arterial, 32, 34
  calculation of, 7–8
  reference ranges for, 3
  venous, 34
sodium (Na). *See also* hypernatremia; hyponatremia
  blood volume changes and, 81
  concentration in blood, 78
  hyponatremia, 82–86
  measurement of, 83
  physiology of, 81–82
  in plasma, 78
  reference ranges, 81
  regulation of, 81–82
  strong ion differences, 7
sodium-potassium ATPase pump, 81
Squires, J. R., 2
Standard Base Excess (SBE), 4. *See also* base excess (BE)
Stewart, Peter A., 2, 5–6
Stowe, Richard, 2
strong ion difference (SID), 2
  chloride regulation and, 95
  description of, 5–7
  examples of, 7
  role of bicarbonate ions, 96
surgery
  calcium monitoring after, 49
  hyperkalemia and, 92
  hypomagnesemia after, 59
sweating
  chloride loss, 95
  sodium levels and, 84
syndrome of inappropriate secretion of antidiuretic hormone (SIADH), 85
syringes
  blood storage in, 36
  filling of, 54
  sample collection, 35, 36, 53

temperature, blood storage and, 36
thalassemias, 14
thiazide diuretics, 84. *See also* diuretics
thirst. *See also* hypodipsia
  absent reflex, 87
  blood volume changes and, 81
  regulation of osmolality, 79
tissue hypoxia, 25, 92
tissue oxygenation, 34–35
total body water (TBW), 77
tourniquets, 94
trauma, 71, 92

urine
  excretion of, 79
  osmolality, 87
  phosphate excretion, 66, 70
  specimen handling, 54
  total calcium in, 54

Van Slyke, Donald, 2
vascular tone, 59
venous blood
   reference ranges for, 3
   sample collection, 35
ventilatory status, 28–29, 34
ventricular tachycardia, 60
vitamin D
   phosphate levels and, 66
   in regulation of calcium, 44
   in regulation of electrolytes, 41
   renal diseases and, 47

VO_2 (oxygen consumption), 8
vomiting
   bicarbonate levels and, 97
   chloride loss and, 95
   hypokalemia and, 90, 91
   hypophosphatemia and, 68

water deficit, 79, 86–88
water load, 79